Beef Cattle

Beef Cattle

Keeping a Small-Scale Herd

BY ANN LARKIN HANSEN

HOBBY
H/F
FARM
PRESS®

An Imprint of BowTie Press®
A Division of BowTie, Inc.
Irvine, California

Barbara Kimmel, *Managing Editor*
Jarelle S. Stein, *Editor*
Jessica Knott, *Production Supervisor*
Honey Winters, *Book Design and Layout*
Lisa Barfield, *Book Design Concept*
Indexed by Melody Englund

Reprint staff:
June Kikuchi, *VP Chief Content Officer*
Karen Julian, *Publishing Coordinator*
Tracy Vogtman, *Production Coordinator*
Jessica Jaensch, *Production Supervisor*
Cindy Kassebaum, *Cover design*

Library of Congress has cataloged an earlier printing as follows:

Hansen, Ann Larkin.
Beef cattle : keeping a small-scale herd for pleasure and profit / by Ann Larkin Hansen.
 p. cm. — (Hobby farms)
Includes index.
ISBN-10: 1-931993-68-8
ISBN-13: 978-1-931993-68-5
1. Beef cattle. 2. Beef cattle—United States. 3. Farms, Small. 4. Farms, Small—United States.
 I. Title. II. Series.
 SF207.H34 2006
 636.2'13—dc22
   2006010532

BowTie Press®
A Division of BowTie, Inc.
3 Burroughs
Irvine, California 92618

Printed and bound in China
13 12 11   6 7 8 9 10

This book is dedicated to my husband, Steve, and our sons, Nick, Phil, and Joe: Thanks for making it possible, and thanks for making it fun.

Table of Contents

Introduction   Why Beef Cattle? 9

Chapter One   Beef Cattle: A Primer 11

Chapter Two   Before You Begin: Fences, Feed, and Facilities 31

Chapter Three   Choosing, Buying, and Bringing Home Cattle 53

Chapter Four   Feeding Beef Cattle 65

Chapter Five   Handling Beef Cattle 87

Chapter Six   Keeping Beef Cattle Healthy 99

Chapter Seven   Breeding Beef Cattle 113

Chapter Eight   Marketing and Processing Your Cattle 131

Acknowledgments 141

Appendix: A Glance at Beef Cattle Afflictions 143

Glossary 149

Resources 155

Index 163

About the Author 168

Why Beef Cattle?

B eef cattle are as much at home on the hobby farm as they are on the range. Adaptable to almost any climate and easy to manage and to market, they are well suited to any farmer with the pasture room and a hankering for a cowboy hat. Although beef cattle require a higher initial investment than any other traditional farm animal, except dairy cows, they require the least amount of daily maintenance.

Attention to the basics of raising beef cattle will reap rewards in the form of a freezer full of homegrown beef as well as extra cash from meat and calf sales. Where cattle are common, so are the auction barns, processing plants, and truckers that make it fairly simple to buy, sell, and process cattle. Americans love beef, so there is a ready market for beef cattle.

There's another benefit to owning beef cattle: they can improve your land. This may come as a surprise, given the reputation cattle have acquired in certain quarters for overgrazing and destroying sensitive lands. But beef cattle are a tool, not a cause. The result depends on how the tool is wielded—just as a hammer can be used to fix a building or to wreck it. Research and on-the-ground experience have demonstrated that, when properly managed, beef cattle can be a highly effective tool for restoring health to damaged grasslands and watersheds. On a hobby farm, well-managed beef can continually increase the richness of your soils, the biodiversity and lushness of your pastures, and the water quality of your ponds and streams.

Beef cattle will also enhance the view from your kitchen window. Every time I look out the window to see our cattle grazing the green slopes of our farm, hear bobolinks singing in our pasture, or prepare homegrown steaks for dinner, I'm glad we have beef!

CHAPTER ONE

Beef Cattle: A Primer

Understanding the basics of cattle evolution, biology, and behavior can provide valuable insight into selecting the right cattle for your farm and caring for your new livestock. Here's a brief history of cattle and an overview of cattle types, breeds, and traits.

OUT OF ASIA, INTO THE WILD WEST

Bos primigenius, the massive ancestor of all modern cattle, stood up to six feet tall at the shoulder and wielded a spectacular set of horns, with a tip-to-tip span of up to ten feet. Herds of these wild prehistoric bovines—or aurochs as they're commonly called—roamed throughout North Africa and most of Europe and Asia, following the melting glaciers of the last Ice Age northward. Between 500,000 and 750,000 years ago, long before humans appeared on the scene, the intimidating aurochs split into two distinctive subtypes, the taurine and indicine (or *typicus* and *indicus*) varieties of cattle. The humped, droopy-eared indicus multiplied in the eastern portions of the aurochs' range, while the humpless taurine, whose ears stand out at the sides rather than flop, spread through the Middle East, northern Africa, and Europe. Their domesticated descendants are the droopy-eared indicus breeds that dominate Asia and are popular in the southern United States as well as the familiar taurine breeds that are common throughout Europe and North America.

Probably because they were so large, well armed, and willing to defend themselves, aurochs were rarely hunted by early humans and were domesticated after goats, sheep, and pigs were. But humans seem to have learned early on to love the taste of beef despite the difficulty of obtaining it. Eventually, they figured out

A pair of bulls carts hay through downtown Weymouth, Nova Scotia, in this old photo. Cattle have historically been used as draft animals as well as sources of meat and leather.

that it was easier to raise, than to hunt, cattle and became ranchers. Cattle bones found in archaeological digs in southwestern Turkey show that between eight thousand and seven thousand years ago, the taurine-type cattle eaten by ancient villagers began decreasing in size, evidence of domestication. These early ranchers clearly selected small, docile animals less likely to charge their owners or disappear in the middle of the night.

While Middle Eastern farmers appear to have been the first to domesticate cattle, farmers elsewhere probably did so not long after. Many researchers believe indicus-type cattle were domesticated separately in Asia. A more controversial theory argues that taurine-type aurochs were also domesticated separately, by North Africans, and later mixed with immigrant indicus to give rise to the distinctive sanga cattle breeds of Africa.

Early Middle Eastern farmers used cattle solely for meat and leather and kept far fewer cattle than they did sheep and goats. When humans began migrating into the heavily forested and much colder regions of northern Europe, they quickly found themselves much more dependent on cattle. On the move north, goats could not pack as many of their owners' possessions on their backs as cattle could, and once people arrived in the new land, neither sheep nor goats were of much use in clearing forests or pulling plows in the heavy soil. As farming spread north, cattle—which are not fussy eaters and not overly bothered by predators and actually seem to like the cold—quickly became the most important livestock species. Oxen (castrated adult male cattle) became northern Europe's primary source of transport and farm power. They remained so until about two hundred years ago, when horses and then the internal combustion engine took their places in most regions.

By three thousand years ago, cattle keepers from Egypt to England had figured out that a cow that pulled a plow or a cart could also furnish milk for the family. Thus the dual-purpose cow and the dairy industry were born. Cows and oxen no longer able to work provided their owners with a large amount of highly nutritious and palatable beef and hides for leather goods.

When English and Dutch settlers brought their cattle to North America in the early 1600s, they adapted their ways of keeping cattle to their spacious new environment. All along the East Coast, cattle not needed for milk or labor were turned into the woods for the summer and rounded up in the fall. Each animal had its ear notched in a way particular to its owner, who had to register his brand with the town officials. In the fall, excess cattle were assembled in herds and a drover hired to walk them to the fresh-meat markets in big towns such as Boston and Philadelphia.

Young Herefords fatten up on alfalfa hay in a California feedlot.

Beef cattle need lots of pasture, and as human settlers multiplied, unclaimed land became scarce. Beef cattle producers pushed farther and farther west in search of the necessary cheap grazing. This made for a longer walk to market in the fall, resulting in skinny cattle. Since fat is what makes beef taste good, a finishing industry grew up near the meat processors, where the cattle could be penned and fattened. As the corn belt of the Midwest developed in the nineteenth century, cattle coming from summer grazing farther west were fattened on excess corn there through the winter and driven farther east to market in the spring. In the 1870s, railroads and then refrigerated railcars reached the Midwest, and the beef processing industry found it more expedient to process cattle in nearby towns, such as Chicago, then ship the finished products to the East Coast, than to ship live cattle.

By the time of the Civil War, the center of beef cattle production—the cow-calf operations that produced the young cattle for finishing—was moving into the

Did You Know?

Domestic cattle, which belong to the genus and species of *Bos taurus*, have no wild siblings. The last known wild aurochs, or *Bos primigenius*, a cow, died in Poland in 1627. Other members of the bovine group are bison and yaks. Sheep, goats, and pigs are more distantly related, belonging to the same family, Bovidae, as cattle.

Near Paisley, Oregon, a cowboy drives a herd from summer range to winter pasture. Cowboys on horseback still move cattle in some parts of the country, especially out west, where the terrain can be rugged and the distances long.

Great Plains, where the herders mounted horses and became cowboys. The cows they herded were Longhorns, left behind by the thousands when their owners, from the Spanish missions and haciendas of Texas and California, left for Mexico after the Texas Revolution of 1836. Most Americans think of western-style ranching as uniquely American, but it really began with the Spanish colonists and their Mexican *vaqueros*, the first cowboys.

During the Civil War, many cowboys and ranchers joined the military and headed east, leaving the cattle to fend for themselves. While eastern herds were decimated by hungry soldiers and civilians, the western Longhorns ran wild and multiplied. When the war ended, there weren't enough cattle left in the East to satisfy demand, let alone supply the wave of soldiers and miners now moving west, and by 1866, beef cattle prices were 200 percent higher than they had been before the war.

Texas rancher Charlie Goodnight returned from the war to find that the 180 head of cattle he'd left behind had grown to 5,000 grazing in the area. Although the cattle weren't worth much

Did You Know?

Historically, the term *cattle* was used to refer to all varieties of four-legged livestock, including horses, goats, and sheep. When referring specifically to bovines, ranchers used the term *neat cattle*. Only in the past 100 to 150 years has the meaning of the word *cattle* changed to refer only to domesticated bovines of the *Bos* genus.

Cattle stampede across the 5,500-year-old cave paintings at Tassili n'Ajjer in Algeria, pull a cart in a 2,600-year-old bas-relief from Nineveh in Mesopotamia, and power a plow in a 1,750-year-old Roman tile mosaic from Saint-Romain-en-Gal in France. Clearly, cattle have been an integral part of human lives, art, and culture in North Africa, the Middle East, Asia, and Europe for a long, long time. Not only have humans used cattle for labor, milk, and meat, but they have also counted the animals as wealth and status and have revered them in myths and religious beliefs.

Ancient Greek and Roman myths often involve cattle. Heracles (Hercules) subdued the great fire-breathing bull of the island of Crete and stole a herd of red cows from the monster Geryon. Crete was famous for bull worship and for being the home of the mythical Minotaur, half-man and half-bull, that lived on human flesh and terrorized a labyrinth under the palace of Knossus.

In India, cattle have been an integral part of daily life and the Hindu religion for several thousand years. A constant presence in ancient art and in rural areas, the sacred cattle of India are traditionally milked and worked, but not eaten, by higher-caste Hindus, who believe many gods and goddesses reside in cows.

In the United States, the myth of Paul Bunyan and Babe the Blue Ox arose from the thousands of loggers who felled the forests of the Northeast and the Midwest, hauled the logs with teams of oxen, and then made up stories around the campfire during long winter evenings.

The tall, silent cowboy trailing his herd into the sunset is one of the great American cultural icons, even though, according to cattle industry historian Jimmy Skaggs, "relatively few of them were the strapping Anglo-Saxon stereotypes of film and fiction. . . . [M]ost of the adult men on trail drives were Mexicans, blacks, and Indians . . . and even a few adventurous young women, some of whom were disguised as young boys." (*The Cattle Trailing Industry*, 1973.)

A cattle owner works in the pasture with his mixed breed herd. Small producers typically have mixed, or commercial, herds, which comprise a variety of breeds and mixed breeds, rather than purebred herds.

in Texas, reports were trickling back of Longhorns bringing an incredible $40 per head in the right market, including the Northwest. In the Northwest, the U.S. Army had forts full of soldiers in need of food and far from any food sources. Goodnight looked up his friend and veteran cattleman Oliver Loving to help, and they rounded up two thousand head of cattle. The pair hired eighteen riders and headed north, edging the Chihuahua desert, crossing rivers, fighting Indians, and turning stampedes. At Fort Sumner in New Mexico Territory, the herd was sold to government provisioners. Goodnight netted the princely sum of $12,000, and the golden age of cowboys and cattle drives was born. Cattle were king.

From that time until the mid 1880s, when a combination of drought and fierce winters nearly wiped out the West's range cattle, one hundred million beef cattle traveled the four great trails: the Goodnight-Loving, the Chisholm, the Shawnee, and the Western. Fortunes were made, legends were written, and the practical beef cow acquired—by way of the cowboy, the cattle drive, and the Hollywood Western— an aura of myth and romance that still warms the hearts of ranchers today.

This mythology has also left many people with the impression that cattle were primarily raised on large ranches. Stepping back from the legend and looking at the figures offers a slightly different picture of the U.S. beef industry. Even at the height of the ranching boom in the early 1880s, the big operations raised only 14.4 percent of the beef cattle sent to market. The majority of beef cattle in the United States always have been, and still are, calved on small farms and ranches into relatively small herds. Small farmers are cowboys, too!

BEEF CATTLE: THE BASICS

For hundreds of years, people have bred cattle to develop characteristics that were best adapted to a particular climate and purpose. Eventually, this resulted in distinctive breeds of cattle, each with a distinctive palette of physical traits. Today, a cattle buyer can choose from a wonderful array of color, build, size, growth rate, and potential meat and milk production to fit cattle to the farm, the climate, and the purposes of the owner.

Although not all cattle are created physically equal, they do share general behavior characteristics. Cattle sense the world differently than we do. They eat different foods and digest them differently. Understanding how cattle operate is key to knowing what to expect from them, what they will like and won't like, and to getting them to do what you want them to do. Understanding and working with cattle's natural behaviors will result in calmer, healthier animals.

BEEF BREEDS

Until the middle of the eighteenth century, cattle were tough, multipurpose animals that were not selectively bred for any specialized purpose. Differences in their sizes, colors, and builds were simply a result of groups' being isolated from one another in remote settlements. Then in 1760, Robert Bakewell, an Englishman, began the first known systematic breeding program to improve the uniformity and appearance of his cattle. The results were published in 1822 in George Coates's *Herd Book of the Shorthorn Breed*, the first formal recognition of a cattle breed. Other breed herd books soon followed, and as the concept of breeding for a specific purpose spread, cattle were divided into two main categories: those bred primarily for milk production and those primarily for beef production. Even the original dual-purpose Shorthorn breed has now been split into Shorthorns for beef and Shorthorns for milking.

More than five hundred breeds of cattle exist in the world today, although only a few are common in the United States. Milking Shorthorn, the ubiquitous black-and-white Holstein, and the rarer Guernsey, Brown Swiss, Ayrshire, and Jersey make up the six primary dairy breeds in the United States. There

Did You Know?

Three breeds of cattle possess the attractive and very distinctive "Oreo cookie" coloring—black except for a broad white band around the middle: the Dutch Belted, the Belted Galloway, and the Buelingo. The breeds differ from each other, however, in their other characteristics and in their uses. The Dutch Belted were a prized dairy breed in the United States until about 1940, while the Belted Galloway, probably descended in part from the Dutch Belted, is primarily a beef breed. The Buelingo is a uniquely American beef breed developed during the 1970s and 1980s by North Dakota rancher Russ Bueling, primarily from Shorthorn and Chianina genetics.

Herefords such as this one have dominated the U.S. cattle industry for more than a hundred years and have a well-deserved reputation for being docile and easy on the feed bill.

are several other minor dairy breeds as well. All dairy breeds produce excess bull calves that are raised for beef, and plenty of beef operations are built on dairy calves.

Hereford cattle, with their familiar white faces, red bodies, and white markings, have been the backbone of the U.S. beef industry since a few decades after their arrival in 1847. Black Angus, first brought to the United States in 1873, are now almost as numerous as Hereford; while that breed's offshoot, Red Angus, established its own breed registry in the mid-1900s. These three breeds, along with the Shorthorn, Scottish Highland, Dexter, Devon, and Galloway breeds, are the major British breeds of beef cattle, so designated because they all originated in England, Scotland, Ireland, or

This sturdy Charolais heifer has a big frame and will produce a lot of beef in a single package.

A long-haired Scottish Highland calf enjoys a good brushing from his owner.

Wales. In general, the British beef breeds are smaller, fatten faster, and are more tolerant of harsh conditions than the continental breeds.

The continental breeds from Europe are generally larger and slower to mature but offer a bigger package of beef to the producer. The most common are the Charolais, the Limousin, and the Saler from France; the Simmental from Switzerland; and the Gelbviehs from Austria and West Germany.

In the United States, new breeds have been developed that tolerate southern heat better than do the European imports. The famous Texas Longhorn, which pretty much developed on its own from Spanish cattle brought over by colonists, provided the starting foundation for American ranching. The American Brahman was developed from indicus-type imports and then was crossed with different European breeds to create the Santa Gertrudis, the Brangus, the Beefmaster, and several other uniquely American cattle breeds.

When well cared for, any breed of cattle will produce good beef. Look for a breed suited to your climate and pasture type and, if the income is important, to the prospective buyer of your beef. Auction barn buyers and finishers will have definite preferences. Most important, get something you like. You may fall in love with the Oreo cookie markings of the Belted Galloway, the shaggy look and big horns of the Scottish Highland, or the gentle disposition of the Hereford.

CATTLE BIOLOGY

Cattle are ruminants, members of a class of grazing animals with four-chambered stomachs adapted to digesting coarse forages other animals cannot utilize. Consequently cattle—as well as sheep, goats, and horses—can make

Of the hundreds of cattle breeds adapted to an enormous range of climates and conditions throughout the world, many are now endangered. The American Livestock Breed Conser-vancy lists eighteen breeds in the United States that need help to survive. On the ALBC "critical" list, defined as breeds that have fewer than 200 registrations each year, are the Ancient White Park (above), the Canadienne, the Dutch Belted, the Florida Cracker, the Kerry, the Milking Devon, the Pineywoods, and the Randall. On the "threatened" list, with fewer than 1,000 U.S. registrations each year, is the Red Poll. The "watch" list, with fewer than 2,500 registrations, includes the once-popular Ayrshire, Guernsey, and Milking Shorthorn, as well as the Belted Galloway, the Dexter, and the Galloway.

use of land too rough, rocky, dry, or wet to grow crops for humans.

Cattle pick their meals by smell and taste, then graze until the first chamber of the stomach, the rumen, is full. Because they have no front upper teeth, just a hard pad, they tear the grass instead of biting it. (This is also why cows don't normally bite people.) Watch a cow grazing, and you'll see her grip a bite of grass between the pad and the lower front incisors, then swing her head a little to rip it off.

The long muscular tongue, as rough as sandpaper, is useful in quickly conveying grass back to the throat. The tongue is also used for grabbing grass, for licking up those last bits of grain, and for a little personal grooming (although cows aren't flexible enough to reach around too far). Copious amounts of saliva—up to fifteen gallons a day for a mature cow—moisten the grass so it slides easily down the throat. Let a bottle-fed calf suck your finger, and you'll be surprised at how strong, rough, and slimy that tongue is.

Once the rumen is full of pasture grass or hay, the cow will lie down in a comfortable spot and, mouthful by mouthful, burp it all back up again. Because she initially swallowed without chewing, the cow now brings those huge rear molars into play and takes the time to grind up the grass into a slimy pulp before swallowing it again, this time into the second stomach chamber, the reticulum. Chewing her cud, as this process is called, takes eight to ten hours each day and involves up to forty thousand jaw movements.

From the reticulum, the cud moves into the omasum and next to the abomasum, the true stomach, then down the intestines. What's not absorbed comes out the back end. Because a cow's diet is high in fiber and fairly low in nutrients, an awful lot comes out the back end, ten or twelve times a day, for a grand total of up to fifty pounds of manure every twenty-four hours. When the grass is young and lush, the manure is almost runny, like cake batter. When cows feed on mature pasture or hay, the manure is drier and more solid.

Along with all undigested organic matter and dead gut bacteria, cow manure often carries the eggs of internal parasites, or "worms," as most people call them. Cows won't graze near their own manure, an evolutionary response to the parasite problem. But cattle show no discretion as to where they poop, so pastures need to be large enough or rotated often enough that the cattle don't foul the grazing areas to the point that nothing is edible.

In addition to the manure deposits, cattle urinate eight to eleven times a day. Both manure and urine are superb fertilizer for pastures. Although the cattle won't graze those areas right away, they will after the deposits decompose.

Having spent the early morning grazing, these cattle now lie around leisurely chewing their cuds, not moving much during the heat of the day. Ruminants usually eat a lot of feed in a hurry, then find a quiet place to rest and digest.

Because cattle need to spend so much time resting and ruminating, they'll graze for only about eight hours a day. (When it's hot, they do much of their grazing at night.) The higher the quality of the pasture or hay, the easier it is for cattle to get enough to eat in those eight hours and to gain weight and bear healthy calves. Young, lush pasture is their favorite food, high in muscle-building and milk-making protein. If it's too young and too lush, however, pasture can cause problems. Cattle digestive systems are set up for lots of fiber, which young pasture and legumes, such as clover and alfalfa, lack. Too much of this type of feed can pack the rumen so

With lush grass to graze on, this cow can produce plenty of milk for her plump calf. Good pasture results in good cattle.

tightly that digestive gases can't escape, and the cow begins to bloat. If the bloat isn't treated quickly, it will put such pressure on the cow's lungs that she won't be able to breathe, and she'll die. For this reason, you should never put cattle on wet or frosted legume pastures and should always provide some dry hay in the spring when pastures are just greening up. Bloat is a fairly common killer of cattle, although it's more common among dairy cattle than beef, due to the much richer diets fed to dairy cattle. Just as in humans, a rich diet seems to create a higher potential for digestive upset.

Older pasture and hay composed of mixed grass and legumes are lower in protein but higher in carbohydrates. This keeps cattle digestive systems in better order, helps fatten the cattle, and keeps them warm in the winter. It is also healthier for pregnant and nursing cows. Mother cows can get too fat on rich pasture, which is hard on their feet and legs and may contribute to difficult calving.

HOOVES AND HIDE

Good feet and legs are important in all cattle. Cows have cloven (two-part) hooves that average around three and a half by four inches. If a cow weighs a thousand pounds, that's a lot of weight coming down on that little hoof every time she takes a step, and she takes a lot of steps in a day to get food and water. Cattle don't like mud or slippery surfaces because they can fall and hurt themselves; they will walk around bad footing if they can. Cattle hooves grow continuously, and long hooves cause lame cattle. If the herd is getting enough exercise, however, their hooves should not get too long. A few rocks in the pasture will help keep hooves worn down.

Cows also use their hooves to scratch themselves and to kick. They can kick both sideways and backward, and they're quick as lightning. If they have horns, they'll also use these to defend themselves, and a horn can do even more damage than a hoof. For this reason, many cattle owners prefer polled, or naturally hornless, cattle. You can also dehorn calves when they are quite young so that their horns never grow.

For pests that are too small to kick, such as biting flies, cattle have long tails for flicking them off, and their thick hides protect them from some insect species. But several kinds of flies can bite through a cowhide, and some will even bore a hole in the hide and lay eggs there. The irritation and discomfort caused by flies can slow weight gains in calves and keep cows miserably bunched on a hot day to minimize skin exposure to the tormenting flies.

To stay warm in cold climates, cattle will grow a longer winter coat. Unlike most dairy cows, beef cows will have hairy udders.

How Cattle Sense the World

Cattle have excellent eyesight, but it works a little differently from human sight. They can see color to some extent, and they see exceptionally well in the

Long hair and a layer of fat keep this steer so well insulated that not enough heat escapes to melt the snow on his back and head.

dark. Because their eyes are spaced so far apart, their horizontal vision (side to side) spans an amazing 300 degrees at a time, with their only blind spot directly behind them. However, their vertical (up and down) range of vision is limited to 60 degrees, which means they have to look down to see where to put their feet when the footing is unfamiliar. What's more, their eyes work somewhat independently of one another, rather than in concert as our eyes do, which gives them poor depth perception. When herding cattle, it is important not to work directly behind them; they can't see you there and will either turn to look at you or spook and run away. When moving cattle into a new area, give them plenty of time to see where to put their feet.

Cattle's sense of hearing is acute, and they can swivel their ears around to hear even better. They dislike loud noises, especially loud sharp noises such as yells from handlers. They get upset and panicky when their handlers start shouting; but they are soothed when you talk to them softly or sing a little. Skip rock and roll, however, and go for soft ballads.

Cattle use quite a bit of verbal communication. They know one another's voices, and they'll learn yours. They'll bellow for feed, bawl for their calves, and moo back when you call them. A cow has a special low moo for when her calf is fed and settled and all's right with the world.

Cattle have a superb sense of smell, which they can use to follow the trail of their calves and to tell different plants apart. They also have a strong sense of taste and, as a result, have strong preferences for some plants over others. Research by Utah State University professor of rangeland science Fred Provenza has demonstrated that calves learn their plant preferences from their mothers and remember them all their lives. Year after year, they'll seek out their favorite grazing spots. We have one small patch of bluegrass that always gets grazed to the ground before anything else is touched, although it looks no different than any other bluegrass!

Biological Traits

Temperature: 101.5 degrees Fahrenheit
Pulse: Adult—60 to 80 beats per minute; calf—100 to 120 beats per minute
Respiration: 10 to 30 breaths per minute
Natural life span: twenty years or more
Adult body weight: Cows—800 to 1,700 pounds; bulls—1,000 to 2,600 pounds
Sexual maturity: Heifers can reach sexual maturity as early as four months but should not be bred until they're at least fourteen months. Bulls should not be used for breeding until at least one year of age.
Heat cycle: Cows will cycle every seventeen to twenty-four days (the average is twenty-one days), with the actual heat (period of time when the cow is capable of becoming pregnant) lasting one to one and a half days.
Gestation: Nine and a half months, or 285 days, give or take a week or two.
Color: Cattle may be black, white, red, all shades of brown, or any combination of those colors.
Coat: Nearly all cattle have short, shiny hair but will grow a somewhat longer, thicker coat in winter. A few breeds, such as the Scottish Highland, are naturally long haired.

CATTLE BEHAVIOR

Although they are not as bright as dogs and cats, cattle are intelligent in their own way. Usually good at taking care of themselves, they'll find a windbreak when it's cold and shade when it's hot, and they'll never forget where the calves are. While cattle learn more slowly than horses, they do quickly learn to come when called if you reward them with

Mother and baby know each other best by smell—the most important "social sense" for cattle.

grain or a new pasture when they get there. When they learn something, they never forget it—especially bad experiences, an important point to remember when you're working cattle. If you are patient and give them time to think things over, they can be taught quite a bit. When introduced to new experiences slowly and quietly, they're also quite curious. If I walk into the pasture and sit down, I'm soon closely surrounded by a ring of sniffing cows, wondering what in the world I'm doing. (If you ever wondered what cow's breath smells like, give this a try.)

Cattle are not democratic. In every herd, there is a hierarchy, from the bossy top cow to the shy one at the bottom of the pecking order that always gets shoved away from the best feed.

Dominance is determined by strength and aggressiveness, but it's all kind of low key. Cows don't fight so much as they test to see who can push whom around. You'll sometimes see a pair of cows head-to-head, hooves planted, just pushing. The winner is the one that pushes hardest and longest. Cows will push around a young bull, too. (Don't worry, he'll get over it.)

Cows have different personalities. One of the delights of having a small herd of cattle is getting to know each cow on an individual level. Some are inherently nervous and will never let you get close. Some are docile and like their heads scratched, while others like to spend most of their time shoving everyone else around. Some are overprotective mothers; others are lackadaisical

Two young bulls engage in a head-to-head pushing contest. Such encounters are a common way for cattle to determine who's going to be higher up in the herd pecking order.

about their calves. Most will nurse only their own calves. But a few will let any calf help herself—kind of like those human mothers we fondly remember from our childhoods who would feed any of the neighborhood kids who happened to be around at dinnertime.

Some behaviors are common to all cattle. Because they are, and always have been, a prey species, cattle are hardwired to run at any sign of danger. Danger to nervous cows could be anything from a strange human in the pasture to a funny smell. They don't stop to think about whether there is really a threat; they just take off. They'll run right over you if they're in a panic, and they panic fairly easily. For animals that are normally fairly slow moving and clumsy, they can move surprisingly fast when frightened or agitated, and they'll jump gates and fences—or trample them.

Cattle would rather run than fight, but they will also fight if they have to. A cow will defend her calf; a bull that decides for some obscure reason that you're competition or a threat will charge you and kill you. Although when kept with the cows beef bulls will usually behave themselves and run away with the cows, they are naturally more aggressive than cows, and they are unpredictable. You can't trust a bull not to charge you instead of running away. Never, ever, turn your back on a bull.

As a prey species, cattle learned long ago that there is safety in numbers. They graze in a group because it's easier for a group to spot and defend itself against predators than it is for a single animal. The herding instinct is not completely dominant, however. Groups of two or three will wander off from the main herd in pursuit of some promising grazing, although usually not too far.

Cattle Benefits

Ranchers discuss some of the benefits of owning cattle.

Gentling the Stream Bank

"I was trained that grazing cattle along streams was wrong. But as I watched, I saw the stream start to change in the fenced-off section. I saw erosion big time. As I watched the rotational grazing along the stream, my eyes told me the cows were gentling the stream bank instead of making it erode. Here, we're seeing some good stream characteristics. There are fifty or sixty different plant species, good root mass, dense growth, and very little erosion."
—Ralph Lentz

Creating Biological Diversity

"We have a vision of a biologically diverse farm, where we raise meat in sync with the environment."
—Juliet Tomkins

Completing the Cycle

"The beef cattle are here to complete the cycle on our farm. For anything to be truly sustainable, you have to have forages and legumes."
—Helen Kees

Teaching Life's Lessons

"I have boys, and I think cattle give them a good example of how life goes on. It's a learning process for the boys. The first calf we got we named Dinner. It's a way of teaching the facts of life."
—Randy Janke

Keeping Everyone Happy

"If you keep your cattle happy, and keep your pastures happy, you'll do pretty well."
—Dave Nesja

Head up and on the alert, this cow is ready to defend her calf if the need arises.

If the pasture is large enough, cows prefer to go off by themselves to calve and may remain away for a day or two before returning with the new baby. Cows also use baby-sitters. Often you'll see a single cow watching over a group of calves while all the other mothers are off grazing and, I presume, gossiping.

THE CATTLE PRODUCTION CYCLE

The life of a beef cow or steer follows a pretty standard pattern in most of the United States. After a calf is born, usually in the spring, the infant stays with the mother until it is weaned (between four and ten months of age). By weaning time, a male calf has been castrated and is ready to go on pasture or hay as a stocker calf, or backgrounder, for several months. An older calf may skip this stage and go directly on feed (presumably this is why all weaned calves are called feeder calves). *On feed* means putting young cattle in a pen instead of a pasture and feeding them a high-protein diet to accelerate growth and fattening. If the steer is an early maturing breed and has been well fed, he may go to slaughter as young as sixteen months of age. If he is a slower-maturing breed or is on a less intensive feeding program, he may be kept until age two or three.

Female, or heifer, calves intended for breeding are kept separate from their mothers and the bull after weaning, usually until they're fifteen months or a little older. At that point, they're bred either by using a live bull or by artificial insemination, usually in mid-summer of the year after they're born. Nine and a half months later, if all goes as planned, they deliver their first calves and officially become cows, instead of heifers.

As long as a cow raises a good calf each year and doesn't exhibit any major personality problems, she's kept in the herd. When she becomes too old or infirm to become pregnant, the owner culls her. Sending a gentle old cow away to slaughter is difficult. I try to console myself with the knowledge that she had a full and happy life on our place and by arranging for her to go somewhere close and quick.

Bulls are usually kept in groups until they're sold as breeding stock. The bull's first calves will be delivered the next year, just ahead of breeding season, so it's safe to use him for a second year. If you keep a bull for a third year, his daughters will be old enough for him to breed. You need to either make other arrangements for the heifers or get a new bull. Bulls can be sold to other cattle owners to give them a couple more happy years in the pasture or sent to slaughter.

This cattle production cycle creates a lot of opportunities to tailor your beef operation to your personal preferences and calendar. A cow-calf operator is on the job year-round, but it doesn't take intensive management to keep cows and calves happy and productive. A backgrounder can buy feeder calves in the fall or the spring and keep them on forages until they're ready for the feedlot. It's quite easy with this system to have cattle only for the summer, so that you don't have to make or buy hay and you can take the winter off. Feedlot operators can work on small amounts of land because they don't need pasture; they do, however, need excellent management skills and a lot of knowledge about cattle nutrition.

Finally, seed stock producers, who raise bulls and heifers for cow-calf operations, must have plenty of experience with breeding high-quality cattle as well as good marketing skills.

This cow's calf has grown fast through the spring and summer and is ready to wean and possibly to sell or even finish at home.

CHAPTER TWO

Before You Begin: Fences, Feed, and Facilities

Y ou *could* stick your new heifers in the garage until you get a fence up around the pasture, but think of the mess! It's much better to have the "three Fs" in place before you bring home the cattle: fences, feed, and facilities.

FENCES

Good fencing is essential to a cattle operation, more important even than shelter. Poor fencing makes for bad neighbors and sleepless nights. If your cattle are constantly in the neighbor's cornfield or causing a traffic hazard on your road, your neighbor and the local sheriff are going to be upset. So you'll need to make sure your fences are heifer high and bull tight.

Well-fed and calm cattle are about the easiest farm animals to fence. Hungry and scared cattle—and those in heat—will jump, break, or trample a weak fence. The fence around your property boundary should hold your cattle no matter what mood they're in. You also need a fence that will discourage them from reaching over or under for a taste of some fragrant plant on the far side and from using the fence as a scratching post. Those sorts of activities break wires and let the herd out for an unscheduled field trip.

Whether building a new fence or rebuilding an existing one, you'll need to pay close attention to the wire gauges, post spacing, and bracing. General guidelines and options for cattle fencing follow. For more detailed fence-building instructions, find a do-it-yourself book or a neighbor who can show you how to build it right. (See Resources.) Fencing projects are best scheduled for early spring while the ground is still soft and the air is cool.

Fence-building tools: Leaning against the wooden post are a post-puller, a post-pounder, and a post-hole digger. In addition to wooden posts, you have the option of metal T-posts for permanent fences, and many types of "step-in" plastic or fiberglass posts for temporary electric fencing. On the ground are wires and tools for electric and barbed wire installation.

OUT WITH THE OLD

Most small cattle operations are started on old farms, and old farms generally come with old fences. If the old fence is still in somewhat good condition, you may be able to get a few more years out of it by running a single electric wire along the inside of the fence to keep the cattle from scratching and leaning on it. If the old fence is half-buried in weeds and strung on rotted or rusted posts, the sooner you can take it down and replace it with something new and tight, the better. Otherwise, you'll be lying awake all night wondering whether this is the night the cows will make a break for it.

Taking down old fencing is a slow job best done in cool weather, when it's comfortable to wear the heavier clothing you'll need to protect yourself against those sharp wire ends and barbs. First, clear away as much brush and as many weeds as necessary to uncover the fence. Then you can start on the fence itself. Take along a bucket, fencing pliers, and heavy leather gloves. You'll also need a post puller, a handy device that looks like a tall jack, which you can borrow from a neighbor or pick up at a farm supply store for $30 to $50.

To disassemble a basic barbed wire fence, start at a corner. With the fencing pliers, remove the metal clip holding the bottom wire to the post and throw it in the bucket. Continue removing all the clips along a couple hundred feet of fence. Go back to the corner and spool up the wire in a big doughnut shape. When the doughnut gets heavy, cut the

To use the post-puller, stand it next to the post, lift the handle, hook the end under a knob on the post, and push down on the handle. Keep pushing until the post is out.

wire with the fencing pliers and lean the doughnut against a fence post for later pickup. Then go back to the corner and do the same with the other wires, working from the bottom up.

After you've removed all the wire, use the post puller to yank out the metal posts. (If they are U-posts instead of T-posts, you may have to rig a wire loop on the puller to make it work.) To pull wooden posts, you can pound a long nail into each, leaving a couple of inches sticking out, then rig a rope or wire loop under the nail and around the post to pull it out with the post puller. Even better, if you can get a four-wheeler, tractor, or car with a trailer hitch close to the post, you can put a loop of chain around the post, then run the chain over the top of a tall board stood on end (next to the post) and down to the trailer hitch. Drive away slowly, and the board will tip over and pull the post up and out.

Immediately fill any holes left by pulled posts to prevent animals and humans from stepping in them and twisting an ankle or breaking a leg. Whichever pulling technique you use, be aware that rotted wooden posts will often break off at ground level, which will save you the trouble of filling the hole, although you may have to do so later when the remaining wood rots. For expediency, I have even sawed off posts at ground level instead of pulling them out.

You can burn old wooden posts, but save any usable metal posts for the new fence. Load up the bent and rusted metal posts and the old wire, and haul them to a junk dealer or a recycler.

IN WITH THE NEW

The two most practical options for a new cattle perimeter fence are high tension and barbed wire. A high-tension fence is the Cadillac of fences, long lasting and presenting a significant physical barrier even to a half-ton cow. It is also more costly than barbed wire. However, even though barbed wire is the cheapest type of fence to build, it will still cost some real money for posts, wire, clips, braces, and a few tools. Per-foot costs for all types of fences change some from year to year, depending on metal prices and where you're located. Your agricultural extension office or fence dealers should be able to give you information on costs so you can budget for your fencing project.

High-tension fence works best where you have long, straight stretches to fence and a decent budget. Because the wire is heavy and stretched very tightly, it requires excellent corner braces and some expertise to install. The high-tensile wire, made extra strong for these fences, is stretched so tightly and anchored so well that tree branches and even cows bounce off after hitting the fence. Sellers of high-tension fencing supplies can furnish installation instructions or put you in touch with installers in your area.

If you're putting up the fence yourself or if you have lots of curves and corners to fence, old-fashioned barbed wire works fine for cattle, although it's not usually recommended for any other type of livestock. Fence installers will build barbed wire fences, too, if you don't have the time to do it yourself. For a perimeter fence, use a minimum of four wires.

High-tension fences are often electrified, adding a psychological barrier to the physical barrier of the wire. An electric fence creates a psychological, rather than a physical, barrier for cattle. An electric fence power unit pulses a static charge through the fence wire. When a cow touches the fence, the charge flows through her to the ground and back to the ground rods attached to the power unit, completing the circuit and giving her a healthy jolt. It's the same effect as shuffling your stockinged feet along a carpet in dry weather to build up a static

Advice from the Farm

The Three Fs

Our experts discuss fences, feed, and facilities.

Fences

"I'm an old-fashioned man, and I still like wire and post versus high tensile and electric. I've built many miles of barbwire fence. I like a minimum of four strands for cattle, and I've gone to longer posts so we can put an extra wire on if we need it. For bulls, we'll put an electric fence in front of the barbwire. Breeds have some differences in how much they'll challenge fences. We've had good luck with our Hereford and Angus bulls."

—*Rudy Erickson*

lifetime, I've always pastured my hay crops, too. The third cutting gets harvested by the cows, and the manure and urine out there were worth a young fortune. It made for a good first crop the next year."

—*Dave Nesja*

Feed

"Utilize what naturally wants to grow on your land, and do what you can to provide an environment for that pasture to be happy. If you've got quack grass, seed in a legume that will enhance the grass. Anything you can do to help what wants to be there naturally is really a good move. Here we've got this heavy ground that's thick with rock and miserable for farming, but it will grow grass. In my

Facilities

"I have a crowding pen and a chute. If you don't mind seeing your animals lie around and die trying to give birth or come in with an eye ruptured from pinkeye, you don't need a facility. Forget the cow for a minute, and think of what having a facility does for the stress on you—it's so much easier. You can have a plan, and you can work your plan."

—*Dave Nesja*

Temporary electric fencing is quickly assembled from step-in posts, plastic electric wire, and a power unit, such as the ones here. Don't forget to ground the power unit to 8-foot copper rods according to the instructions that come with the unit, and keep the tester handy for tracking down problems.

charge, then touching your kid brother to give him an electric shock. When done correctly and well maintained, an electric fence is extremely effective for subdividing pastures and keeping groups of cattle separate.

Always use smooth wire for electric fences; it is illegal in many areas to electrify barbed wire. To augment your barbed wire with an electric barrier, you can use offset insulators to mount a smooth electric wire along a barbed wire fence. This is a good combination, especially if you're keeping a bull separate from heifers.

The standard low-tension, soft-wire permanent electric fence uses permanent posts and works well with three wires. The top and bottom wires carry the charge, and the middle wire is a grounded wire. This is necessary for those times when the ground is covered with snow or is very dry and acts as an insulator rather than the receiving end of the circuit. A cow sticking her head through the fence will connect a hot wire with a ground wire and get a jolt.

Portable electric fencing, built with lightweight step-in posts and usually a single strand of plastic wire, is used during the grazing season to temporarily subdivide pastures into paddocks. This fence technology is slowly revolutionizing grazing in this country. For relatively little time and money, livestock owners can subdivide pastures into paddocks to manage their grazing, which improves pasture growth. You can set up and take down portable electric fencing almost as fast as you can walk because all it requires is stepping a line of posts into the ground, then unreeling the wire and popping it into the clips on the posts.

Cattle are quick to figure out when an electric fence is not working and will walk through it if the grazing looks better on the other side, so check electric fencing often. It's also critical to exactly follow instructions for sizing and installing the power unit and for grounding it properly with a series of copper rods.

If your cattle are unfamiliar with electric fencing, you'll need to train them to recognize and respect it. To introduce them to the concept, run a temporary electric wire along the inside of a wooden corral or a holding pen fence. Out of curiosity, the cattle will see the wire and sniff it, giving themselves a jolt on the most sensitive part of their anatomy—the nose. Because the solid fence is in front of them, they'll back away rather than jump forward at the shock. Once they've figured out what the wire means, take it down. You don't want it there when you're working cattle because if one should touch it accidentally, you'd have some upset cattle on your hands in uncomfortably close quarters. If you have cows that know about electric fences with new calves that don't, you needn't worry. The calves will learn without any special training.

Any fences enclosing a confined area where cattle might be crowded or stressed from handling—such as corrals and holding pens—should be made of heavy-duty wood or metal. These fences should be high enough that cattle won't even think about jumping them: at least five and a half feet for small, calm cattle and six feet or higher for large cattle or cattle unused to people or to being handled. Build these fences low to the ground, too, so your cattle won't try to scramble underneath. It's amazing how small an opening a cow will try to get through when she's frantic.

All pasture fences should be at least four feet high and have a wire close enough to the ground to keep calves from scrambling underneath but not so low that you can't trim the grass under the wire. (I put the bottom wire a foot off the ground.) When building a new fence, make sure to leave enough room to use a brush mower or weed trimmer on both sides, and avoid installing the

In a holding pen, a high fence that also provides a visual barrier is ideal. The salt and mineral feeder built into the fence gets cattle in the habit of being in the pen and is also handy for filling.

fence on steep banks and close to rock piles and big trees. Keeping fences clear of vegetation at least doubles their life-spans and makes the inevitable repairs much easier to do.

All fences need gates for moving cattle, people, and equipment in and out of pastures and pens. In your corral and handling facility area, gates should be solid metal or wood and bolted on so a steer can't stick his nose under the bottom rail and flip it off the hinges. Metal and wooden gates are wonderful in perimeter fences, too, but if your budget doesn't allow for as many nice gates as you would like, you can build a "poor man's gate" by extending the fence wires across the gate opening. Instead of attaching them to the post at the far side, attach them to a four-foot stick. Put a wire loop at the top and bottom of the gatepost. To close the gate, insert the ends of the stick into the wire loops.

Did You Know?

Henry W. Vaughn, in his 1949 *Types and Market Classes of Live Stock*, notes that cattle have some special advantages over other types of livestock. They can feed on materials that other animals won't eat, such as corn stalks and coarse hay, and they can be pastured on low, wet land not suitable for other animals or on rough and arid land not usable for crops. He also notes that cattle prices tend to have less seasonal fluctuation than prices of any other type of livestock.

Lay out your fences and gates for ease in moving your cattle. It's much easier to move cattle through a gate in a corner than through one in the middle of a long, straight stretch of fence. Field gates are usually best located in the corner nearest the watering or grain feeding area because your cattle will be moving back and forth between water and pasture regularly. It's hard for them to understand when they have to walk away from the direction they want to go, then double back because you put the gate in the far corner.

FEED

Cattle eat a lot. They need big stretches of grazing and lots and lots of hay when they can't graze. Unless they're being fattened for slaughter or the hay is of very low quality, adult beef cattle don't need grain, but giving them a little grain each day keeps them tame and can be used to train them to come when called. Before you bring your cattle home, you'll need to have edible pasture for the growing season and edible hay for the winter. Having some rolled corn or a grain mix on hand to help you make friends with your new livestock is a good idea, too.

Below is a general discussion of pastures, hay, and grain. You can find out the specifics of what is usual and available in your area by talking with your neighbors, the staff at the feed store, or your agricultural extension agent. A word of caution when asking people for advice in those areas having a lot of dairy cows: often the advice you get on feed and everything

Beef Cattle Facts

The United States is the largest producer of beef in the world and home to a hundred million head of cattle. About thirty-three million of those are beef cows, and fourteen million are young stock being fattened for slaughter. Another nine million are dairy cows, each of whom produces a calf annually; half of those calves are male and are raised for beef.

Although beef is big business in the United States, three-quarters of all beef cattle spend at least part of their lives on small farms. An estimated 830,000 farms and ranches in the United States keep beef cattle, most in small herds. Here in Wisconsin, the average beef herd is just over twenty cows, which means there are a few big operators and an awful lot of us small, part-time beef producers.

About half of all U.S. beef cows live on the rangeland and pasture between the Mississippi River and the Rocky Mountains. The rest are scattered throughout the United States, and there aren't a lot of counties in this country that don't have some beef cattle. Dairy cattle are concentrated in the states around the Great Lakes, in the Northeast, and in areas of the Pacific and Rocky Mountain states.

While beef cows and calves are widely distributed through the country, cattle being fattened for slaughter are much more concentrated. Eighty percent of cattle in feedlots are in Colorado, Kansas, Nebraska, Oklahoma, and Texas, where there's plenty of room and easy access to feed grains. Many small feedlots are still scattered through the rest of the country, and most small operators at least fatten a steer at home every year or two for their own freezers and often fatten a few to sell to friends and neighbors.

else will be aimed at milking cows and will not be well suited for beef cattle. Be sure to ask specifically for information on beef cattle, as their needs are different from those of dairy cattle.

PASTURE

Your pasture is the centerpiece of your beef operation. It normally makes up the bulk of a herd's diet, and cattle that feed on good pasture are healthy and happy. Providing good pasture also means not having to provide as much hay, and the less time and effort you invest in hay, the more likely it is that your beef operation will wind up in the black at the end of the year. So it's worth your while to take the time to figure out which plants and soils you have now and to plan for what you want to have.

What You Have

Start by estimating how many cattle your pasture can sustain. You can do this by first calling your extension agent and asking how many acres of pasture it takes to support a steer or cow in your area. (This could be anywhere from one and a half to forty acres, depending on whether you live in the humid Southeast or a semidesert area in the West.) With this figure, you can then calculate how many head of cattle you can theoretically carry on your land. Keep in mind, however, that this is just an estimate. The actual number will vary considerably depending on the fertility of your soil, whether it's a dry or wet year, and whether you have uplands, lowlands, or something in between. Keep your carrying capacity estimate on the conservative side, at least until you have a few years' experience under your belt.

Once you have a rough idea of how many cattle your land may be able to support, take a walk in your pasture. What's growing there? Grasses and legumes, such as clover and alfalfa, that cattle will thrive on, or weeds, such as burdock, goldenrod, milkweed, and thistle? A weed, in this context, is not necessarily a bad plant; it's just something that cattle won't eat. If these weeds are taking up a lot of space in your pasture, you don't have as much feed as your acreage might indicate. Is there a lot of old vegetation that was never grazed off and is now choking new growth? Are there bare spots? These conditions reduce available pasture, too. Revise your estimate of the pasture's cattle carrying capacity, if necessary, based on the condition of your pasture.

Taking these two simple steps will tell you how many head of cattle you can start with. Again, it's best to start with a conservative estimate, allowing for more acreage per cow or steer than you think you'll need. It's cheaper and less hassle to be long on feed and short on cattle than the other way around.

What You Want

Once you know what kind of pastureland you have, you can determine what you want and how to go about achieving that. Ideally, the majority of the pas-

On arid rangelands with native bunch grasses and sagebrush, you may need ten to fifty acres to feed a cow and calf for a month.

ture would consist of palatable grasses, with a healthy component of legumes. Weeds and bare spots would be scarce or nonexistent. Achieving this happy state may take a few years of managed grazing, mowing, and fertilization. You may want to add plant species by overseeding, that is, scattering seed in an established pasture. Much of the fertilization and all of the grazing will come from your cattle. Your job is to manage the cattle so they do a good job of fertilizing and grazing.

If you have an overgrown pasture, the quickest way to jump-start it, especially in regions where there's enough rain, is to mow it. This will stimulate fresh grass and clover growth. Weed control may have to be geared to the type of weeds you have, although in the majority of cases regular mowing will keep most weed species to tolerable levels. Species-specific advice on weeds is available through your extension office, on some university databases, and in many farm publications, as well as on the Internet. (See Resources.)

If you're turning old corn and soybean fields into pastures, talk to your extension agent or your seed dealer about what to plant. Generally, you'll want a combination of grasses and legumes suited to your soil type that will start growing early in the spring and last well into the fall. For legumes, I like alfalfa with its long taproot where our soil is dry and sandy and red clover in wetter areas or clay soils. Since I often use the pastures to make hay, I prefer grass species that are tall and don't grow in hard-to-mow clumps. Finally, there are the plants you get for

This man is testing the soil at a farm to determine its composition and decide what steps the farmer can take to improve the land.

free. Species such as quack grass will appear on their own in your pastures in many regions. Often that's OK. Quack grass may be a weed in the lawn, but it's good eating for cattle.

Whether you're renovating old pastures or creating new ones, it's very helpful to get a soil test. Some extension services offer soil testing, or you can check with your local seed dealer for phone numbers and addresses of soil testing laboratories in your area. The test results will indicate whether you need lime or fertilizer, and then you can proceed accordingly. Make certain you specify that you are testing for pasture because soil amendment and fertilizer recommendations are calculated differently for row crops.

In dry areas where soils crust over, the best method for renovating pasture is to break up the crust either mechanically or with heavy animal trampling, then fertilize and plant. When animals are used to break the soil cap, they also fertilize the soil with their manure. In arid regions, careful grazing management is critical to maintaining good pasture, requiring a combination of short periods of intense grazing followed by adequate rest periods. Of course, this contradicts the traditional advice of putting only a few cattle on huge tracts of rangeland or pasture for extended periods. But research and experience have shown that rotational grazing reinvigorates pastures in arid areas more effectively than any other method, while extended low-impact grazing generally makes this type of pasture far less productive in the long term.

A final note on pastures: it's worthwhile to patrol them for old bits of wire, stray nails, and other metal garbage before the cattle arrive. Cattle will eat this stuff, and occasionally it will perforate their stomachs and make them ill. This is called hardware disease, and it's far better to avoid it than to treat and attempt to cure it.

HAY

When pastures are dormant due to drought or cold, your cattle will need hay. Ideally, the hay should be a mix of grass and legumes cut before they go to seed and baled after they're thoroughly dry. While edible, hay made from old vegetation is not very palatable. Good hay has a green tinge and smells pleasant. Hay that has been rained on after it was cut and before it was baled will be more brownish and have a distinctive smell, but if it's not moldy, it will work fine for beef cattle. Hay that has become moldy because it was baled too wet can make animals sick.

Of all types of livestock, beef cattle can best tolerate the lowest-quality hay, but if it's really bad, they won't eat as much as they need to and will lose weight. Pure alfalfa or clover hay is a little rich for beef cows—too high in protein and too low in carbohydrates to generate the body heat needed in cold weather as well as mixed hay will. (You can leave it for the dairy cows, which need the protein for milk production.) There is one exception: alfalfa hay can be an important component of a fattening ration for animals getting close to slaughter.

Mixed grass-alfalfa hay is great feed for cattle, especially when it's green, such as that shown here, and sweet smelling.

Calculating how much hay you will need for the nongrowing season can be a little complicated. A pregnant cow will need more hay than a bull or a young steer. If the weather is warmer than usual, the cattle will eat less hay. If it's colder or windy, they'll eat a lot more. A fat cow will need less hay to keep warm than a thin cow will. If it's low-quality hay, the cattle may need supplements (usually grain, but other products, such as brewery and ethanol by-product, turnips, and silage, work) to meet their nutritional requirements. As a rule of thumb, an average 1,000- to 1,200-pound beef cow will need a ton to a ton and a half of hay to get her through a five- to six-month winter-feeding period. That's not too exact (this is not an exact science!), but it will give you a starting point. Ask your neighbors how much hay they buy for a cow for the winter or an extension agent how much he or she recommends buying, then round up. It's always better to have too much hay than too little.

If you have only a few head of cattle, it may make sense to buy your hay and have it delivered. Hay is cheapest in early summer, after the first cutting of the year has been made, and most expensive in late winter when supplies are shortest. Keep track of hay prices through the area farm paper or local feed store to make sure you're paying a fair price. Inspect the hay before you buy it to make certain it's of decent quality and not moldy.

Hay comes packaged in small square bales, large square bales, small round bales, and large round bales. If you buy from a dealer, you'll generally pay by the pound or the ton. If you buy from a neighbor, you'll often pay by the bale. Large round bales are usually the cheapest to make or buy on a per-pound basis, and small squares are usually the most expensive. What type of bale you buy will depend on what's available in your area and how you intend to store it and feed it to your livestock.

Large and small square bales deteriorate quickly when exposed to the elements, so you'll need a roofed shed or a hayloft for storage. Round bales can be stored outside. In a wet year, the top few inches of round bales will get moldy, but the cattle will push that aside to get to the good stuff underneath.

Feeding is quicker and less frequent with large bales, whether square

Making Hay

Making hay, to oversimplify things, involves first mowing the field and letting the cutting lie until it's dry, then raking it into rows, and then picking up the rows with the baler, which compresses the hay into bales and wraps them with twine. In our area, with good, dry weather, this takes three or four days from start to finish. On our place, I cut and rake the hay and hire a custom operator to bale it into large round bales. Then I put the hayfork on the tractor and move the big round bales off the field into a neat line along the fence.

A farmer feeds small square bales directly on clean snow. When fed on a daily basis, the cattle will clean most of it up before they lie in it or foul it with manure.

or round, but you'll need a tractor with a hayfork to move them. Small squares are easily moved and fed by hand, so if you don't have equipment for moving bales, small squares make the most sense. You can feed bales directly on the ground, but be aware that cattle love to play with their hay, and they will throw it around, lie in it, poop on it, and generally waste a third to half of it. Save yourself a lot of money and mess and build or buy a hay feeder. Several types of metal feeders are available at farm supply stores.

When you have more than a few head of cattle, it's expensive to buy hay. The alternatives are to have a neighbor make hay for you on your land or to make it yourself. Traditionally, if a neighbor makes your hay for you, he gets half. You are responsible for fertilizing and reseeding the hayfield when necessary. Another option is to hire a custom operator, who will make your hay for a per-acre or per-bale fee. Making your own hay (see "Making Hay") will require a fair investment in equipment, but doing so allows you to work on your own schedule.

Grain

Unless you have the equipment and the experience, it's easier and usually cheaper to buy grain than to grow it yourself. Talk to the folks at your local feed store to find out what your options are for types of grain and mixes. Corn is generally the preferred feed. I have it rolled to break up the kernels, which enables the cows to digest more of the grain rather than just passing it out in the manure. I also add one-third oats to two-thirds corn to lower the calorie count (our girls are plump enough) and because older farmers I know believe oats are good for digestion and make cattle's coats shine.

Build or buy a rodent-proof storage bin for your grain. Well, maybe that's a little optimistic; it's incredible what small openings rats and mice can crawl through. But try. As a few cost-effective examples, you can use metal garbage cans, build a wooden lid for an old metal water tank, or construct a plywood bin inside a barn or a shed. Just don't store your grain on the floor in the sacks it comes in—that's asking for major rodent problems.

Handling Facilities

Cows are bigger, stronger, and faster than we are, so when it's time for vacci-nations or shipping, we need a way to make them hold still (other than trying to rope them or grab a tail). You can't hold a half-ton cow if she doesn't want to be held, and generally she doesn't. That's why cattle owners should—for their own safety, if nothing else—have a cattle-handling facility. Of course, if you're handy with a lariat, enjoy a rodeo, and don't place a high value on unbro-ken bones, you can get by without one, but your veterinarian won't like coming to your place.

Cattle facilities have four parts: a corral or holding pen, a crowding tub, an alley, and a chute with a headgate. You can add pens, add a Y in the alley for sorting, and even add hydraulics to the chute for lifting cattle, but the small oper-ator really needs only the four basic components. Even if you just have a

Young calves crowd around a grain feeder. Calves quickly learn that grain tastes great.

Poisonous Plants

Poisonous plants lurk in most pastures, but fortunately cattle usually know better than to eat any. All the same, it pays to walk your pastures regularly, keeping an eye peeled for potential problems.

No matter where you live, chances are that some plants in your pasture could poison your cattle. Nationwide, many dozen plant species are known to cause illness and death in livestock. Fortunately, most (though not all) poisonous plants taste icky. If your cattle have enough to eat, they probably won't touch anything that's bad for them. But if your pastures are stressed by drought or have been heavily treated with nitrogen fertilizer, or if it's very early in the spring and the only plant that's green is also poisonous, you should be alert for problems. Most poisonous plants (not all, however) must be consumed in large quantities to cause toxicity in cattle. Cattle can even develop immunity to some poi-sonous plants, but don't count on that.

Six different classes of poisons have been identified in the various plants, the most important being the alkaloid and glycoside groups. Alkaloids affect the nervous system, causing loss of motor control, bizarre behavior, and death. Jimsonweed, a common species of the western United States, is probably the best known example of a plant that kills with an alkaloid poison. Glycosides basically cause death by suffocating cells. The animal is breathing, but the oxygen in the bloodstream is blocked from being transported into the individual cells. The buttercup, which brightens low pastures in early spring, is a familiar glycoside-containing plant.

Other familiar plants that are dangerous to cattle include black locust, black nightshade, bracken fern, castor bean, curly dock, death camas, dogbane, horsetail, locoweed, lupine, milkweed (several species, but not all), oleander, pigweed, and tobacco. White snakeroot, common throughout the Midwest, causes the "trembles" in cattle and can kill humans that drink milk from cows grazing it. Thousands of settlers in the Midwest died of milk sickness in the early 1800s, including Nancy Hanks Lincoln, Abraham Lincoln's mother.

For more information, see the Resources section or contact your local agricultural extension agent.

This squeeze chute and headgate at the end of a narrow running chute are on cement for a good anchor and to keep them out of the mud. The sides of this chute can be moved inward to better immobilize an animal.

couple steers for the summer, you'll still need some sort of pen and a chute to load them onto the truck at the end of the year. Make sure to locate your facilities where they will allow easy access for a truck and trailer.

The pen or corral holds the cattle until you're ready to work them. A few at a time are herded into the crowding tub, a small, circular pen with a swing gate that pushes them into the alley. The alley is narrow so cattle don't have room to turn and must go single file. At the end of the alley, a grate is raised to let a single animal at a time into the chute, where the cow or steer's neck is caught in the headgate. This holds the cattle in one place so you can safely administer shots, put on ear tags, or do

whatever else may need to be done. Once the vaccinating, breeding, or castrating is done, the headgate is released, and the animal moves back out to pasture. The headgate is then reset for the next patient.

When planning your handling facilities, follow these few basic principles. First, unless it never rains in your area, put your facility inside a building or on cement, or both. It's no fun working in the mud. Old dairy barns can be easily converted for working beef cattle, as can nearly any sort of shed. Our facility is on a cement pad outside the old barn.

Second, plan for curves, good footing, and good lighting. A slippery surface will make cattle nervous, and they will balk if they see a barrier ahead.

They will, however, follow a curve around to its end. They also will move more readily toward a well-lit area than they will a dark one.

Finally, if you're building with raw materials rather than buying manufactured components, follow the recommended dimensions exactly. The recommended alley width for your breed of cattle may seem incredibly narrow, but anything wider and you'll have cattle trying to turn around, which can result in injuries. Never force a heavily pregnant cow into the alley or chute— she might get stuck.

There are many plans for cattle facilities available (see Resources). Most of them are much more elaborate than necessary for a small operation of fewer than thirty head of cattle. With a little creativity, it's possible to build a fine facility at a reasonable cost in a fairly small area. Ours is built on a 75-by-50-foot concrete pad next to the old dairy barn. We also use the area for moving tractors and equipment in and out of a shed. So we hung a lot of gates that we usually leave open for mechanical traffic but close when handling cattle in order to subdivide the area into three separate pens that wrap around to feed into each other and then into the crowding tub. Since I'm no carpentry whiz, I bought a manufactured tub, alleyway, and headgate, all of which we bolted to heavy-duty wooden posts that we cemented in so the cows couldn't push things out of alignment. The result may not be aesthetically beautiful, but for working cattle, it functions wonderfully. The water tank and grain feeders are also situated on the concrete inside the

Shade isn't mandatory in cooler climates, but cattle appreciate having it on hot days.

holding pens so the cattle go in there to eat and drink all the time and are very familiar with the setup, which makes for calm and easy handling.

You, too, will be glad you went to the trouble of putting together a solid handling facility the first time you need to load or vaccinate cattle. A good facility turns what could have been a dirty, dangerous, nerve-wracking, half-a-day chore into a fairly pleasant hour with the herd.

SHELTER, WATER, SALT, AND MINERALS

Once you've got the fences built, hay in storage, and a handling facility ready, there are only three tasks left to do before you're ready to bring home your cattle: provide shelter, water, and salt and minerals. Cattle tolerate an amazing range of weather conditions and do fine outside year-round in nearly all climates. There are a few times, however, when they should have some sort of shelter. In the "ice belt" (those states between the snowbelt and the no-snowbelt), where there can be months-long stretches of damp, almost-freezing weather, cattle are healthier and happier if they can get out of the mud and the wet. It doesn't have to be anything fancy; an old shed open on one side or an old barn bedded with old hay, straw, or sawdust will work. However, you'll need to borrow or rent a skid steer with a bucket to clean the place out in the spring.

Cold winds can be hard on cattle, too. If you can arrange a place where they can get out of the wind, they'll do much better. This can be anything from a belt of trees to the side of a building. Extremely hot weather is also tough on most types of cattle, so having shade available, again either under trees or near a building, makes a big difference to them.

There are two must-haves for raising beef cattle: a water tank and a salt and mineral feeder. Without salt and minerals in their diet, cattle will become ill. Water, of course, is essential for all animals, and cattle should have a constant supply of fresh water.

The water tank should be big enough that cattle can't push it around. Either metal or plastic tubs work; both are available at farm supply stores. I've also seen water tanks made out of old truck tires and out of used milk bulk tanks from dairy farms. It's best to put the water tank on a cement pad to avoid having a mud hole around it. Keep the tank full with a garden hose, or in warm weather, hook the hose up to a float valve in the tank and it will keep itself full.

Ponds, lakes, streams, and rivers can be good sources of water—until the cows start pooping in them and turning the banks into mud, which they always do if you give them free access. This can destroy the pond or stream. However, if you're rotationally grazing so that the cows are near the water for only short periods, cattle are generally beneficial for the pond, lake, stream, or river. Alternatively, you can build a water access, a narrow

A Scottish Highland calf cools off by taking a dip in a pond. Your cattle's access to natural water sources should be limited to short periods of time so the water isn't fouled.

concrete ramp that allows for one or two cows to drink at a time but keeps them out of the water and off the banks.

You can either build or buy a salt and mineral feeder. Install the salt and mineral feeder where the animals will have free access to it at all times— along a fence line, attached to the side of a building, or on a free-standing post. Make sure it has a roof to keep out rain and snow. The feeder should have two compartments: one for pure salt and one for a mineral mix formulated specifically for your region. Both are available at farm supply stores or your local feed dealer.

CHAPTER THREE

Choosing, Buying, and Bringing Home Cattle

With fences, feed, facilities, shelter, water tank, and salt and mineral feeder in place, you're ready to go shopping for cattle. By this point, you've invested more time and money than you would have for any other type of farm animal, except dairy cows, but it will all pay off in cattle that stay home, eat well, and handle easily. From now on, it gets a little easier, and you'll have less daily work to do than you would for sheep, goats, pigs, or poultry.

You have a few options for where to buy cattle and several choices in what kind of cattle you buy. Fit your purchase to your budget, the size of your pasture, how much time you'll have each day for chores, and whether you're interested in beef for your freezer or in building a herd.

Whatever age or sex you buy, the minimum number of cattle you should purchase is two. Cattle are herd animals and hate being alone. They will adopt a goat or a donkey or anything else handy as a companion, but they thrive best when in the company of their own kind. If you're buying a steer primarily for your own consumption, keep in mind that most families will take a year or two to eat a single steer. So plan on selling the extra steer at the auction barn or the extra half or quarter of beef to friends or relatives.

Following are some general guidelines for choosing and purchasing cattle.

WHAT TO BUY

Steer calves purchased in the fall will be ready for butchering when they are between sixteen and thirty months of age, depending on the breed and your feeding program. This means that a male calf bought after fall weaning could be ready

These well-grown steer calves have been moved into the weaning pen and are figuring out the water tank in their new home.

for the processor as soon as the following fall. Heifer calves bought after fall weaning will be ready to breed the following summer, provided they grow well through the winter and spring.

Any calves you buy should have been weaned for at least three weeks. They should also have received a nine-way vaccine, then a booster two to four weeks later. Heifers should be wearing the small metal ear tag that shows they've had a brucellosis vaccination. Bend down and look behind steer calves to make sure the castration got both testicles, or you could have a bull on your hands by mistake.

If you want cattle around just for the summer, you can buy steers in the spring and sell them in the fall. If you're buying for your own freezer or to sell as finished (ready to slaughter) cattle, they should be started on some grain right away. They'll quickly learn to come running in from the pasture for their daily grain ration.

If you're buying breeding stock— whether cows, heifers, or a bull—finding high-quality cattle is more important than if you're raising cattle for processing. Start the search early and take the time to find out about different breeders. If you're planning on showing cattle or enrolling your kids in the 4-H beef program, look for operations with good show records. If your objective is to get a decent cow-calf herd started, it's more important to find sellers with calm, clean, reasonably good-looking cattle.

Another option is to buy dairy bull calves. They will produce fine beef.

However, it takes a lot more grain to fatten them up, the cuts aren't as nicely shaped, and there's a smaller proportion of meat to bone and by-products. Since only cows give milk, bull calves are a waste product in a dairy operation and are generally sold sometime between three days and a few weeks old. Consequently, they aren't weaned, so you'll have to bottle-feed them milk until their digestive systems are mature enough to handle grain and forage. Although they're a lot of extra work, these calves are available year-round, are very inexpensive compared with all other cattle, and can be transported in the back of a van.

Occasionally, you may be able to buy an orphaned beef calf or one that's been rejected by her mother. Some dairy farmers breed their heifers to a beef bull for easy calving with their first calf, and these half-beef, half-dairy calves are a bargain. Treat orphans or dairy-beef cross calves just as you would a dairy calf.

WHAT TO LOOK FOR IN CATTLE

The only way to acquire an eye for good cattle is to look at a lot of cattle. It takes a few years to develop that eye, but there are some things even a beginner can spot if you take your time and know what to look for. I'm always a little nervous when buying cattle, and I forget things if I hurry so I make a mental checklist. Cattle, too, get a little nervous when a stranger shows up in their pasture or pen. Give them time to settle down again. Lean on the fence or stand quietly in the pasture, and take a good long look at their shape and how they act.

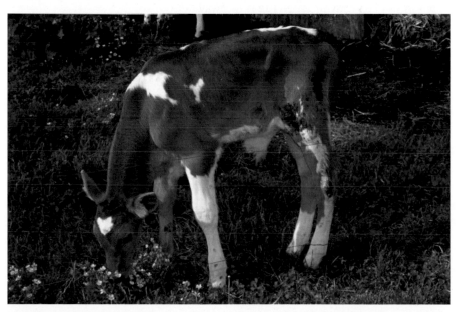

This Guernsey bull calf, even though it's a dairy breed, will produce excellent beef and is a lot cheaper than buying a weaned beef calf.

Advice from the Farm

Buyer Beware
Our experts offer some advice on buying beef cattle.

Have Patience
"We found someone who was quite willing to sell us two bred cows, for $800 apiece. Hindsight is so clear; I believe now that we paid top dollar for old cull cows. However, the cows were bred, and the sellers were very helpful in guiding us in the care and feeding of the cows and in newborn calf care. The bad part of all this is that within one year we had to butcher both of these cows due to rectal prolapse. With both cows gone and two bull calves left, we bought two nice heifer calves, which is what we should have done in the first place but were too impatient. We now have had four cows for several years, which has been a good number for us. We carry twelve animals over the summer—four cows, four of last year's calves, and four newborn calves—and eight over the winter, after the four one-and-half-year-olds are butchered."
—Linda Peterson

Check Looks and Temperament
"I like them so they look straight all the way back on their backbone. There's almost an edge to them. They look upright, they look strong; it's almost an eager look. I always check the feet—some cattle, it seems like a hereditary thing, have really long hooves. Temperament is a big thing for me, too, with little kids around. I've got one cow

with a little bit of attitude, but she's never offered to come after you. I've got another one that's so meek and mild she always gets pushed back."
—Randy Janke

Narrow the Focus
"There's so much going on now with numbers—expected progeny differences, relative this and that—and it's so far extended that the average young person needs to just focus on one or two things when choosing breeding stock. The ones I strongly recommend are calving ease and temperament. You need to stick with cattle that won't explode every time you come around."
—Rudy Erickson.

Know the Seller
"I've never bought from a sale barn. We bought private treaty or from a neighbor's auction where we knew the herd."
—Dave Nesja

A prospective purchase should be looked over carefully for conformation and good health before buying. A good place to start developing your eye for cattle is at a county fair like this one, where the judges will often explain the good and bad points to the audience, such as the low-hanging udder of this cow.

SHAPE

Whatever the breed, beef cattle should ideally look thick and square, like a big hairy rectangular box on legs. While a dairy cow will look like a wedge, with the narrow part at the front, a beef cow should be blocky. The back should be straight and the line of the belly nearly so, not tapering too sharply up to the hind legs, with a rib cage that is rounded, not flat. The hindquarters should look broad and meaty, especially in steers and bulls, and the legs should be straight. Hooves should be even and short, definitely not so long that they curl upward, and they should point straight ahead.

Steers should have thick necks and fleshy forequarters. Cows and heifers should be slimmer through their shoulders and necks and have more feminine heads. Cows should have high, well-shaped udders. Both sexes should have

Watch Them Move

When buying cattle, watch how they move and hold themselves. You should observe no lameness or hunched backs when they walk. Cattle that won't relax, that keep their heads high and bodies braced, may be wild and hard to handle. Cattle that won't let you anywhere near them to look them over could be a problem, too, but don't expect to walk up and pet them either. Unless they're show cattle, most beef cattle aren't accustomed to being approached too closely by strangers.

The telltale dried feces on this calf's rump indicate coccidiosis. It's common and curable but not something you want to bring home.

a wide muzzle, indicating that they can take big bites of grass.

People who show cattle generally prefer longer legs, while producers who rely heavily on grazing often prefer short-legged animals.

HEALTH

Look closely for any signs of illness or discomfort. Eyes should be clear, not mattered (crusted) or inflamed.

In summer, coats should be smooth and shiny (except for long-coated breeds, such as Scottish Highland). A dull, flat-looking coat generally means internal parasites or poor nutrition. In late fall, winter, and spring, the coat should be uniform and thick. Bare spots could mean ringworm; although rarely a huge problem, it is a chore to treat.

Pay special attention to hooves and legs, checking for growths or swelling,

especially at the top of the hooves and at the joints. There should be no swelling on the jaw, neck, shoulders, or brisket.

An occasional cough in adult cattle usually isn't anything to worry about (cows do get colds and runny noses), but constant coughing signals problems. I'd hesitate before bringing any coughing animal home to my herd. In young calves, constant coughing or labored breathing may indicate pneumonia, a dangerous condition.

Check the hind ends of calves to make sure they aren't matted with manure. Manure that is more water than feces is typical of scours, another common and dangerous affliction of young calves.

Polled cattle are naturally hornless, but if you're buying a horned breed that has been dehorned, make sure the spot is well healed and not sprouting any more horn.

HOW CATTLE ARE PRICED AND SOLD

The pricing and availability of cattle generally follow a yearly cycle, which varies by the age and sex of the animals. Calves and steers are usually sold by weight, while heifers and cows for breeding are sold either by weight or by what the market will bear.

Bulls are normally priced according to their quality or what the owner thinks he or she can get. Bulls are most expensive in the spring and early summer, when they're in high demand for the breeding season, and cheaper in the fall.

Bulls are the most expensive category of cattle to buy, and they will have a huge effect on the quality of your herd. Choose carefully!

Feeder calves—those that have been weaned and are ready to go on pasture or on a finishing ration—are usually least expensive in late fall, when the market is flooded with calves born the previous spring that are being sold before winter. Stockers or backgrounders (feeder calves headed for a few months on pasture before going on a finishing ration) are expensive in the spring, when landowners are buying cattle to keep their pastures grazed during the growing season.

Open, or unbred, heifers will be most expensive in spring, just before breeding season; the price tapers off through fall, when they're cheap because it's not economical to winter an open heifer. In addition, a heifer that didn't "settle," or get bred, during the summer may be infertile and good only for being finished and slaughtered. Bred cows will be most expensive in the spring, just before calving, and cheaper in the fall, when you will have to feed them through the winter. Before buying a bred cow, always have a veterinarian check to see whether she is really pregnant.

Cattle prices are listed, usually weekly, in local and state farm papers. If you have a farm radio station in your area, prices are typically announced daily or weekly, and some state extension services list current cattle prices on their Web sites. Cattle prices are listed either as dollar and cents per pound or as dollars and cents per hundredweight. If, for example, I were looking for feeder calves, I would look down the column until I came to that category under the listing for the auction barn closest to my farm, and then I'd start with the midrange, 400- to 600-pound category. If steer feeder calves were listed at $1.00, then the price for a 500-pound feeder

Body Condition Scoring

Body condition scoring (BCS) is a simple method for assessing how fat or thin cattle are, which will help when you're looking at cattle to buy or assessing the cattle you already have. It's kind of like learning to play a guitar: Anyone can quickly figure out a few simple chords, but getting good at it takes some practice. So it is with body condition scoring: it's a simple system and you can start using it right away, but getting exact results takes experience.

While dairy cattle owners mostly use a five-point scale for BCS, beef cattle owners use a nine-point scale (see right). Although specific criteria are used in BCS, it's a fairly subjective measurement. The point of doing it is to assess whether the cattle are in the right condition to raise big calves, to make it through a cold winter, or to breed back easily. Cows that are too thin produce less milk, which results in smaller calves. Thin cows also need more feed in the winter, since they don't have a lot of insulating fat to hold body heat and so have to work harder to stay warm. Cows that are too thin may also have trouble getting pregnant, while cattle that are too fat will bring a lower price at auction due to all the weight that's essentially a waste product.

The optimum BCS for a cow is 5 to 7, or "pleasantly plump." A cow's body condition should be scored three to four months before calving, allowing time to adjust the ration and to let them gain or lose enough weight to be in optimum condition for calving and rebreeding.

To score cattle, look at the fat covering the ribs, shoulders, back, and tail-

head. The numbers break down as follows:

1. Emaciated: no fat visible anywhere; ribs, hip bones, and backbone clearly visible through hide
2. Poor: spine doesn't stick out quite so much
3. Thin: ribs visible; spine rounded rather than sharp
4. Borderline: individual ribs not obvious; some fat over ribs and hip bones
5. Moderate: generally good overall appearance; fat over the ribs and on either side of tailhead
6. High moderate: obvious fat over ribs and around tailhead
7. Good: quite plump; some fat bulges (pones) possibly visible
8. Unquestionably fat: large fat deposits over ribs and around tailhead; pones obvious
9. Extremely fat: tailhead and hips buried in fat; pones protruding

Jewel, our herd matriarch, is in the 8 to 9 BCS range and moves a little slower than the rest of the girls. Even though she's really fat, she produces an excellent calf each spring and breeds back easily. I could separate her from the rest of the cows and put her on a diet, but since she's happy and doing her job, I haven't.

steer would be $500. Keeping track of prices gives you a good idea of what you should be paying when you buy, but keep in mind that it may be worthwhile to pay a little extra for cattle you know are healthy, have been vaccinated, and come from good parents.

WHERE TO BUY CATTLE

Finding cattle for sale is a matter of checking ads in local newspapers or regional farm papers; looking at bulletin boards at the feed store, farm supply store, and rural gas stations; and just asking around. If there's a beef producers association in your area, join it. If you don't know whether there is one near you, give your extension agent a call and ask. An association is a great place to network and get some background information on area beef cattle operations and auction barns. County and state fairs are other good places to find beef producers with cattle for sale. Go to the cattle shows, walk through the barns, and visit with the exhibitors. Two additional sources for leads on cattle for sale are your local artificial insemination service and veterinarian.

Auction barns move a lot of cattle, but they're no place for beginners to buy. If you go, take a friend who is a good judge of cattle and can help you avoid the ones that are sick, are wild, or have bad hooves and legs. You may want to make a few dry runs to the barn, going early to visit the pens and then watching the auction without buying, to give you a feel for how the bidding process works and how cattle are moved in and out of trailers, pens, and the auction ring.

A better idea is to buy cattle directly from a seed stock producer or a commercial producer. Seed stock producers raise purebred cattle for sale as breeding stock and are good sources of quality animals. Commercial producers generally have mixed herds of several breeds or crossbred cattle being raised for beef production instead of breeding stock. These won't be registered purebreds, but often they're of good quality and reasonably priced; sometimes they aren't. Most commercial cow-calf operators sell their calves after weaning in

This good-looking Charolais-Limousin cross calf glows with health as it waits its turn for showing at a county fair. Fairs can be excellent places to meet other cattle people and get leads on animals available for purchase.

the fall, and this can be an excellent opportunity to purchase.

Dairy bull calves are a common item at auction barns, but it's better to buy directly from the farmer and save the calf the stress of being hauled twice to strange places and exposing him to who-knows-what bacteria and disease at the auction barn. When you buy directly from the farmer, you can make sure the calf is at least three days old and has received colostrum, his mother's immunity-boosting first milk. This is critical to a calf's health.

Bringing Cattle Home

Once you've bought your cattle, you have to get them home. If you're buying from a breeder, he or she may be willing to deliver the cattle for an extra fee. Otherwise, you'll either have to hire a cattle hauler or do it yourself. There are usually haulers for hire in any area where there's cattle, and you can track one down by asking neighbors or calling a local auction barn. If your only option, or the option you prefer, is to transport your new livestock yourself, you'll have to buy, borrow, or rent a trailer (unless you're buying small calves you can fit into a pickup or small truck).

When the cattle arrive at your farm, ideally you'll turn them into a solidly fenced small pen or barnyard, with water available and some nice hay scattered around. Don't rush them out of the trailer; give them time to look around and step down carefully. Of course, they may decide to all come out in a rush, but let that be their decision. Once they've had a few hours to get a drink, find the salt, and get a bellyful of hay, open the pasture gate. By then, they should be calm enough to walk, not run, out. They might start grazing immediately or go on a tour to figure out where the fences are.

If you're using electric fencing and the cattle you've bought are familiar with it, you can turn them out with no worries. If they don't know what an electric fence is, you'll need to train them as outlined in the fence section of chapter two. Don't try to train them as soon as they get off the truck, however. That's a lot to ask of already-stressed animals and may send them over the fence and back toward their previous home. You can have the wire ready in the pen, but don't turn it on until they've settled down.

You can also turn new cattle directly into the pasture. If you do this, plan on spending some time watching them to make sure they don't charge and break fences or decide to hop over and head back where they came from. Make sure they find the water, salt, and mineral within twenty-four hours.

Calves in a Van

Our first beef cattle were dairy calves. To bring them home, we put a tarp down in the back of the van with some straw on top, and I had the kids sit with the calves and keep them lying down. They won't make a mess if they're lying down!

Two calves try to figure out a supplement feeder. Cattle are naturally curious, and if you keep them calm, they will quickly adapt to new surroundings and new routines.

Watch your new cattle particularly closely for the first two or three weeks. Are they grazing contentedly, bunched tightly, or spending a lot of time walking instead of chewing? If they're walking all the time, you've got pasture that's too poor, and you should give them some supplemental hay. If they're bunched, you probably have a fly problem and should provide a shady area for them to get away from the worst of the flies. You may also need to put up some sort of rub—a rope or padded post impregnated with fly repellent—that will put the repellent on the cattle when they scratch themselves. Are the cattle spending plenty of time lying down and chewing their cud, or are they always standing up and acting nervous? If they aren't lying down, something is bothering them, and you'll need to figure out what it is and fix it. Once, we had a bear stroll through the back pasture, and the whole herd went through the fence! We moved the cattle to a paddock close to the house, where the dogs could keep the bears at a distance and the cattle could chew their cuds in peace.

Make sure, too, that your cattle are drinking enough water. In temperate, reasonably dry weather, they'll come for a drink at least once, and usually twice, a day. If they're not drinking and it's not raining, there's something wrong with your water setup. I remember one cold winter day when our cattle wouldn't drink, and I found out why when I touched the water. It was carrying an electric charge from a shorted-out heater!

CHAPTER FOUR

Feeding Beef Cattle

A griculotural colleges teach entire courses on cattle feeding and nutrition, but for the small-scale beef producer it doesn't have to be that complicated. There are three basic components to the cattle diet on a small farm: pasture, hay, and grain. This chapter will help you determine which foods to feed at which times and under what circumstances, in the way that's best suited to your cattle, your land, and your budget.

One important rule to remember is: whenever you change your cattle's diet—whether when moving the cow herd from hay to pasture each spring or when moving steers on to a finishing ration—do it slowly. The naturally occurring bacteria in cattle digestive systems, which transform food into nutrients, need time to gear up for a new ration.

PASTURE

Being able to tell a good pasture from a bad one is easy once you know what to look for. Converting a bad pasture into a good one is also fairly simple, if you do a little planning and use fences to control where the cattle graze and for how long. The most critical ingredient in the recipe for developing and maintaining a high-quality pasture is, as the old saying goes, "the footsteps of the owner." A daily stroll across the pasture will give you a feel for which plant species grow through the season, how your cattle graze, and where the rich spots and poor spots are. That knowledge is the basis of good pasture, and good pasture is the foundation of a healthy ecosystem on your farm and of healthy cattle. Many graziers record what they planted in each pasture and how many days cattle could graze there each season.

PASTURE QUALITY

The grasses and clovers that cattle like to eat grow differently from trees, shrubs, and some weeds. If you understand this difference, you'll understand why mowing and grazing are the keys to good pastures. Grasses and clovers have a "growing point" at or near the ground. When a cow bites off a blade of grass or a clover stem, the plant quickly regrows from this growing point. Trees, shrubs, and weeds, however, grow from the tips of their branches and leaves. If a person with pruning shears or a cow nips off this type of plant, it takes a long time for it to regrow because it's lost that special set of cells programmed to grow the plant. That's why, when you prune a shrub, it stays pruned for months. By contrast, you have to mow the lawn every week—the cutting actually stimulates it to grow faster by removing the older leaves that are getting in the way of the growing point at the base of the plant. Grazing has the same effect on individual plants as mowing does, so grazing, when correctly managed, results in lush pastures.

Unmanaged grazing, however, can devastate a pasture. This is because when a mower or a cow shears off the leafy part of the plant it temporarily depletes the food supply to the roots, and some of those roots die. Dead roots put a lot of organic matter into the soil, which is great for holding water and keeping the soil moist, but a great many live, healthy roots are necessary for a thick, lush pasture. You want a balance between dead roots and live roots. If you cut your grass every day or let your cows graze the same plants every day, you kill too much of the root, and the grass will become stunted or even die. If the process goes on too long, the soil loses much of its plant cover and becomes vulnerable to air and water erosion.

Weeds are especially abundant with extensive or continuous grazing systems, in which the cattle feed in the same pasture for an entire growing season. Because the cows keep the grass and clover so short, the weeds have no real competition for sun or water and thus can grow with little restraint. In the spring, when all the plants in an extensive pasture get off to an even start and are growing like gang-

Did You Know?

Cows are not picky eaters. An amazing variety of feeds are used in commercial feedlot finishing rations, starting with grains ranging from corn, barley, oats, rye, and millet to an eclectic collection of additives derived from other crops and from food processing by-products. These range from beet pulp, wheat bran, molasses, and potatoes to cottonseed hulls, linseed meal, soybean meal, and brewery by-products. Latest on the feedstuffs scene is a by-product from corn-based ethanol production, which is cheap if you live near an ethanol plant but has a short shelf life when the weather is warm. Cattle's willingness to eat such a wide variety of feeds is one of the main reasons they have adapted so well to so many different environments.

Green pasture on good soil provides all the nutrients beef cows need.

busters, an extensive pasture looks great. By late summer, when the rain has slacked off, the spring growth spurt is over, and the cattle have kept their favorite plants short, a lot of these pastures are full of big weeds, tiny grass plants, and skinny cattle.

By contrast, grass that isn't grazed while it's still fairly young and tender gets stiff from hard-to-digest cellulose as it matures. The tall grass blades shade the growing point near the soil, and growth slows or stops. Some older plants in a pasture are OK; cows need fiber in their diet, just as we do, and older pasture supplies it. However, the older the plant is, the slower it grows, and the less palatable it is to cows. Keep in mind, too, that a certain amount of old growth left over the winter can protect roots and growing points from freeze-thaw cycles that heave the soil and break roots. Too much, though, and the ground will be shaded and slow to warm

in the spring, and new growth will have a tough time struggling through the old stuff to reach sunlight.

In summary, a thick pasture full of grasses and legumes that cattle like (as discussed in chapter three) and lacking the weeds they dislike—with grass that isn't too old or too short—is ideal for the health and growth of cattle. This type of pasture provides the added advantages of growing longer into dry spells, greening up sooner in the spring, and staying green longer in the fall, which means money in your pocket that you won't have to spend on extra hay.

ROTATIONAL VERSUS EXTENSIVE GRAZING

When all of these factors are taken into account, it would seem clear that the best way to graze cattle is to mimic how you mow your lawn: let them graze an area thoroughly for a short period, then put them somewhere else while the area rests and regrows. This is called rotational or management intensive grazing. Since the science of it was worked out in the 1960s and 1970s by Allan Savory, founder of Holistic Management International, and a host of other researchers, farmers, and ranchers around the world—and portable fencing became readily available in the 1980s—rotational grazing has been quietly revolutionizing pasture and range management.

Yet extensive grazing is still by far the most common pasture system in the United States. It can work pretty well for a cow-calf operation. If you have a lot of land and just a few head of cattle, it may be the most economical choice. The cattle can be turned out for the grazing season and left largely on their own; you only need to make sure they have water, salt, and a mineral mix on hand at all times. If the pasture is big enough, no one will starve, although if it gets dry in the late summer and grass growth stalls, the cattle will need hay.

The biggest long-term problem with extensive grazing is that it can wreak havoc with the soil, water, and vegetation. Cattle that return to the same areas day after day to graze or rest will kill the plants and compact the soil. If there's a stream or pond in the pasture, they'll trample the banks into mud. With no lush vegetation to shade the water and no roots to hold the soil, the water temperature will rise and the banks erode, clouding the water and silting up the bottom. This is devastating for many aquatic species, especially prized ones such as rainbow trout. When cattle rest in the shade of the same trees day after day, the trampling can destroy the delicate feeder root systems and kill the trees. When palatable grass and clover plants are constantly

This range grass is still green and high in protein—ideal feed for these young cows. The greatest single factor in determining the quality of grass is its stage of maturity.

Cattle in a pond or stream look picturesque—but it's hard on the water quality. Limit cattle access to natural waterways by fencing off a small drinking area, and leave the rest of the bank to recover its vegetation.

grazed into the ground, noxious weeds can turn into major problems. Most of us have seen pastures that seem to consist mostly of thistles or spotted knapweed, where all remaining grass is kept grazed to the height of a golf course green. Especially in dry climates, where even without cattle it's difficult for vegetation to prosper, extensive grazing can be devastating.

Setting up rotational grazing for beef cattle is fairly simple. Using whatever type of fence you prefer, split your pasture into several paddocks. Step-in posts and plastic electric wire are the cheapest, quickest, and most common choices for paddock fences. If, however, you don't care for the constant maintenance required by electric fencing, you can put up permanent paddock divisions.

Paddocks should be sized to provide enough pasture to feed the herd for at least three days but usually no longer than a week. A shorter period tends to make beef cows too fat, and a longer stretch allows them to regraze plants that are just beginning to regrow. If you're grazing steers with the goal of putting on as much weight as quickly as possible, you can shorten the rotation time to twenty-four, or even twelve, hours, although it's not necessary if the pasture is in good condition.

Paddocks need to be rested anywhere from a couple weeks during the spring flush of growth in high rainfall areas to several months in hot, dry regions. Getting it all right takes some experimentation, talking to other rotational grazers in your area, and practice. Fortunately, rotational grazing is a forgiving process, and the cattle will probably do fine while you fine-tune your system.

Keep offering hay in the spring until grass growth can keep up with grazing. The cattle will let you know when they don't need hay—they'll quit eating it.

Portable fencing makes it possible to change paddock sizes and configurations at the drop of a hat. In those years when I have more cows, I subdivide the land into more paddocks and graze them for shorter periods. When I have fewer cows, I cut back on paddock numbers and don't graze as tightly. Some years, I mess up or the rains just don't cooperate, and I have to feed hay in August. Most years, I also graze all or part of our hayfields, either early in the spring or late in the fall or during a dry stretch when the pastures have given out. Anytime a cow can harvest forage for herself, rather than your having to do it mechanically, will save you time and money.

Overall, rotationally grazing pastures produces significantly more grass as well as more palatable grass than extensive grazing does. In practical terms, that means faster-growing animals and fewer out-of-pocket costs for feed. Organic matter in the soil increases, which helps hold moisture in. Consequently, grass growth continues longer into a dry spell. The pasture gets thicker and lusher. Trees are hardier and quit dying from soil compaction. The tall grass along streams shades the water, keeping the temperature at a more optimal level. The roots hold the soil of the stream banks tightly—preventing the stream from eating away the banks, depositing dirt on the stream bottom, and widening the stream and making it shallower. As a result, the stream stays narrow and deep, and the water is clear.

GRAZING MANAGEMENT BASICS

Some grazing management practices are dependent upon climate. In arid areas, according to Allan Savory, a high density of grazing cattle is necessary to break up the soil crust and to work seed as well as fertilizing manure and urine into the ground. In the Deep South, where high summer temperatures prohibit grass growth, some cattle owners plant warm-season annual forages for grazing when pastures aren't producing. In our area, the Midwest, the major concern is weed control. I mow paddocks once or twice a season, just after they've been grazed. In general, it takes grasses and clovers about three days to begin regrowing after they've been grazed. You want to mow within that time window so you're only mowing plants the cattle didn't graze, not cutting regrowth. In the areas where the ground is too rocky or steep to mow, I hand-cut weeds, preferably before they go to seed. Because we're a small operation and I like being out in the pasture in the evenings, I find this to be a pleasant chore.

Every two or three years, it's a good idea to test your pasture soil. This involves taking a small shovel and a bucket and gathering samples from the top few inches of soil at several locations in the pasture. Mix up the samples, put some in a plastic bag, and send them to the soil testing laboratory. In a few weeks, you'll get back a report showing nutrient levels for nitrogen, potassium, and phosphorus, as well as the pH level (acid-alkaline). When you submit your samples, ask the lab to test for trace minerals, too, especially calcium. If your pasture is deficient in any nutrients or minerals, you'll need to purchase and spread the correct amendments or hire someone to do it for you. To find a soil lab in your area, to get precise directions on how to take a sample, or to locate lime and fertilizer dealers, start by asking your local agricultural extension agent or the feed and seed dealer. Ask what time of year is best for taking samples and spreading amendments in your area, too, because this makes a difference in the accuracy of the test and in maximizing the benefit from anything you add to the soil.

Cattle can make good use of western land that is too steep and dry to farm. When the grazing is well managed, cattle can help restore and maintain the land's healthy state.

Finally, pastures should not be monocultures (limited to one plant type). You don't want to eat the same thing for every meal, and neither do your cattle. A mix of several types of grasses, a few different legumes, and an eclectic selection of other plants, such as dandelions and plantains, will furnish a nicely balanced diet for cattle. Most pastures are already very diverse, but if you're short on one of those plant categories, you can buy the seed and work it into the pasture. How you work the seed into the soil and at which time of year depends on the type of seed and the condition of your pasture as well as where you live. Ask your seed dealer, extension agent, and any other specialist you can find for advice, then use what seems to fit your situation.

HAY

As mentioned in chapter two, beef cows are at the bottom of the list when it comes to needing top-quality hay. As long as the hay isn't moldy or nothing but stems with no leaves, it'll do for beef cows. Pure, high-quality alfalfa hay, which is high in milk-producing protein and low in body heat–producing carbohydrates, can be harder on beef cows than old, coarse hay. You might want to use it if you're fattening steers during the winter. Knowing when to buy or harvest hay saves you money, and planning your winter feeding to fit your equipment and schedule can save a lot of time and labor.

WHEN TO BUY OR HARVEST

If you're buying hay for cows and calves, you don't need the expensive stuff. During the growing season, keep a close eye on the weather in your region. Dry seasons quickly create hay shortages, driving prices sky high. Buying early and in quantity during good years and storing the excess is generally your best bet. In covered storage, hay will last for years, and even round bales stored outside will last two or three years if the bales are tight and kept on dry ground. Don't pay for any hay until you've dug into a few bales and checked for mold, weeds, and stem content.

If you're having hay made for you on your land or making it yourself, don't be in too much of a hurry. Waiting a little longer into the season to make your first cutting of hay gives you several advantages: The hay will be taller, giving you more volume. It will be higher in carbohydrates and lower in protein, which is good for keeping cattle warm in the winter. The weather in most areas will be more settled, with

---

Did You Know?

In his *Farm Book*, compiled from 1774 to 1826, Thomas Jefferson reports, "In fattening cattle they will eat from $\frac{1}{3}$ to $\frac{1}{4}$ of their weight in turnips per day besides hay. They will fatten in four months on turnips and hay alone, or in three months on a change of food. They prefer carrots to turnips. Lots of them will suffice and fatten faster."

Large round bales are widely available and, if you have the handling equipment, are an economical and easy way to feed your herd.

less likelihood of a surprise rainstorm ruining the cutting. Grassland nesting birds will have a better chance of getting their babies fledged and out of the nest before the hay mower comes through. In addition, if you're depending on someone else to make your hay, he'll likely be less busy later in the season. Here in northern Wisconsin, our dairy farming neighbors like to have their first hay cutting done by late May or early June. I generally wait until late June to cut ours. If I want to slow it down even more, I can graze the cattle on the hayfields in early spring to set the crop back a week or two.

Our local dairy farmers also like to take three or four cuttings off their hayfields each year. I take two—I'd rather have the cows harvest it by grazing a couple times than have to haul out the tractor and Haybine again. This extends the grazing season and cuts my out-of-pocket feed costs. As an added benefit, the cattle fertilize as they graze.

WINTER HAY FEEDING

How you winter feed your cattle depends on your setup and preferences. If you have just a few head of cattle and are using small square bales, it's easy to construct a wooden hay feeder or buy a metal feed bunker. If you're feeding round bales, you'll need round bale feeders sized for the bales you have. These are widely available at farm stores, and the pieces can be hauled in a small trailer for assembly at home. Some local stores might even deliver.

If you've got a shed for the cattle, you can feed hay inside. This is nice in foul weather, but it greatly increases the volume of manure you'll have to clean up next spring because cattle like to stick close to the hay in winter. If you feed outside, you can either feed in the same spot every day or keep moving the hay feeder to a new location.

If you're feeding in the same location every day, the happiest situation is having the hay feeders on a cement pad.

These round bales were placed and fenced off on a poor area of pasture before the snow flew. As the round bale feeders are moved back week by week through the winter, the cattle leave a layer of mixed hay and manure on the ground—a superb fertilizer that will reinvigorate the pasture without my having to do any manure spreading.

This eliminates the mud and simplifies cleanup in the spring. If you don't have a cement pad, your dirt feeding area will become a "sacrifice area," so churned up by the cattle that it's unlikely you'll have anything growing there for a long time. Put a sacrifice area where it won't destroy the view from the kitchen window. In either case, plan on scraping the feeding area or the shed clean in the spring and spreading the moist mix of manure and old hay on your pasture or hayfields with a manure spreader. This typically requires a Bobcat with a bucket, a tractor, and a manure spreader. In some areas, custom operators will come in and do the job for you, or you can borrow or buy your own equipment. If you don't clean up the area, chances are that in most areas of the country you'll have a terrific infestation of stable flies, as the manure-hay mixture is optimal for their breeding. Stable flies will keep cattle running around the pasture to get away from their ferocious bites or miserably bunched in a corner instead of grazing.

The other option for winter hay feeding is "outwintering," or feeding hay on pasture away from the barn. The feeders are moved each time they're emptied. The manure and wasted hay is spread as the cattle are feeding, and there's no spring cleanup, except perhaps for running a harrow or disk over the area to break up the big clots and spread it a little more evenly. Be aware, however, that in areas where winters are wet and the ground doesn't freeze this system will quickly make a huge muddy mess in pastures. Where the ground freezes, it's a terrific way to renovate poor spots in pastures. Because it's frozen, the sod won't be badly cut up by the cattle, and in the spring, the ground will be covered with a layer of manure

Advice from the Farm

Feed for Cattle

Here are a few words from our experts on feeding your cattle.

Wintering

"We have no barn. The round bales are stored outside. The cattle are wintered in a half-wooded horseshoe 'coulee' of about forty acres, with a year-round spring. We feed the round bales right on the ground, and the cows clean it up pretty good. What little they do not eat is their bedding. No sense in having the cows eat all the hay and then haul in straw to a barn just to have a damp place where the sun doesn't shine and disease builds up."

—Mike Hanley

Going Light on the Grain

"I probably have them on a finishing ration four to seven months, because I don't feed a lot of grain. I put them in a little early, and I don't feed them really heavily, maybe fifteen to twenty pounds of a mix of corn and barley. The rest is silage and hay. I think you get a better meat and fewer health problems."

—Donna Foster

Doing It by the Pail

"For finishing at home, doing it by the pail method—buying your corn and oats and protein supplement and mixing it yourself—saves the extra expense of having the mill mix it. There's a lot of supplements out there, and you have to get one you're willing to work with. A lot contain antibiotics and stuff, and we don't use those."

—Rudy Erickson

Putting Up Very Good Feed

"Consider putting up oat hay, sweet clover, sorghum-sudan hybrid mixes, forages mixed with grains, and soybean hay. All these are things I've seen put up in my lifetime, and with the new equipment today it can be done a lot easier than in the old days. Any of these fed properly to livestock is very good feed."

—Dave Nesja

This yearling steer was started on grain before weaning last fall and now in the spring comes eagerly for his daily ration inside a converted pigsty. Starting grain feeding early teaches cattle to come when called and is an easy way to fatten a homegrown steer for market.

mixed with hay, the best fertilizer there is. The cattle won't graze the area until late in the following growing season, but the pasture there will be lush and deep green for years to come.

With outwintering, you can either haul hay out to the feeders a few times a week, or you can set all your bales out during the fall and not have to start a tractor all winter. This is what I and quite a few other beef and dairy producers do in our area: In October, I calculate how many bales I think I'll need for the number of cattle I'm carrying through the winter and for the length of time I think the ground will be frozen. With a tractor and hayfork, I set out the

round bales in three or four rows, spaced fifteen to twenty feet apart on all sides. When the bales are in place, I cut and pull off all the twine (or netting wrap) by hand. I build a three-strand electric fence around three sides of the rows, leaving one side open. I roll the round-bale feeders out of storage and pop them over the bales at the head of the rows on the side that I left open. Then I stick plastic step-in posts into the next row of bales and run two strands of electric wire across them to keep the cows from the rest of the hay.

Constructing this "hay corral" takes me twenty to thirty hours, but it's pleasant work in moderate weather and it means I won't have to start the tractor on cold mornings or fight with twines or round bales frozen to the ground. Instead, once or twice a week I unplug the electric fence, walk out to the hay corral, move the portable electric wire back one row, tip the feeders on their sides, roll them over to the new row of bales, then walk back and plug in the fence again. This takes all of fifteen minutes. What's left of the old bales is bedding for the cows, keeping them clean and out of the snow. In the spring, what was a poor area of pasture will be fertilized, and by late summer, the pasture will be deep green and growing taller and lusher than anywhere else.

GRAIN

Cattle love grain, and that's a good thing because it brings them running when we want them and can produce

Feed Additives and Growth Supplements

More than 90 percent of cattle being fattened for slaughter are implanted with hormones and given antibiotics in their feed to make them grow more quickly and to convert feed to muscle more efficiently. When done correctly, the economics of these practices are persuasive. Properly used hormone implants will add forty to fifty pounds of weight to a finished steer for a couple bucks' investment, while antibiotics and ionophores (a particular class of antibiotics with a different mode of chemical action in the body) increase the efficiency of digestion by 10 to 20 percent, which results in quicker weight gain. Antibiotics in the feed at low levels also help prevent illness and disease in what is (in a feedlot) a very crowded and dirty environment for cattle, which are being fed an unnatural diet.

Hormone implants are placed in the middle third of the ear and are either a synthetic estrogen that will increase muscle gain, a synthetic androgen that decreases protein breakdown and so increases muscle mass, or a combination of the two. A wide variety of dosages and mixes is available so a feedlot manager can gear an implant program precisely to the type and age of cattle being fattened. However, implants inserted incorrectly or the wrong implant type for the animal can diminish the benefit. An estimated 25 percent of implants have abscesses around the implant site, reducing their effectiveness, and implants in larger breeds can result in animals so large that processors will discount the price. Implants can also increase toughness and delay marbling—the fat deposited inside the muscle—and so lengthen the time the animal must be on feed to reach a higher quality grade.

Concern over the effects of these practices on the environment and humans has grown markedly in the past few years. An article by Michael Pollan in the March 31, 2002, edition of the *New York Times Sunday Magazine*, titled "This Steer's Life," probably did more than any other single event to galvanize public interest in what had been fed to the beef they were eating. Mr. Pollan revealed that most antibiotics sold in the United States end up in animal feed, not in people, and thereby add considerable impetus to the ominous and accelerating development of antibiotic-resistant human infectious agents. The synthetic hormones that are slowly released into cattle bloodstreams through implants leave measurable residues in both the meat and the manure. Manure that gets washed into nearby streams and lakes, in turn, leaves measurable levels of hormones in the water, where fish with abnormal sexual characteristics have been found. Some scientists believe, Mr. Pollan wrote, that this build-up of hormonal compounds in the environment may be connected to falling sperm counts in human males and premature maturity in human females.

These yearling steers on excellent pasture will fatten easily without a lot of grain.

tender, tasty beef from even mediocre animals. Although the most common grain fed to cattle is field corn, they will eat a wide variety of other offerings, from oats and wheat middlings to soybean meal and cottonseed meal. For the small operator, corn is usually the cheapest, most available, and easiest grain to buy in small quantities (under a ton). Corn should be ground or rolled so the cattle can digest it better.

Talk to your feed store about other feed options in your area or additional additives, especially if you are fattening a steer for slaughter. Corn alone isn't high enough in protein to satisfactorily fatten an animal during a short period. Corn and good pasture will do the job, but if you don't have lush pasture or high-quality legume hay during the finishing period, you should talk to your feed dealer about formulating a finishing ration. In addition, in some regions there are cheaper alternatives to corn, making it worth your while to inquire.

Beef cows whose purpose is to produce calves, not meat, don't need grain if they're on good pasture in summer and adequate hay in the winter. But I give them a little anyway, as do most beef producers I know. It's called "training grain." A small amount—a pound or less for each cow—brings them running every morning when I call. Because our bunker feeder is inside the holding pen, getting them in when I need to work with them is never a problem. As a bonus, the cows teach their new calves about grain each year so when it's time to start the calves on grain, they know exactly what to do.

You should start calves on grain no later than when you wean them. If you want to start them before weaning, set up a "creep feeder" that will keep the cows out of the calves' grain. A creep feeder is a pen or shed with an opening too narrow for cows but wide enough for calves. Inside is a bunker feeder for the calves. I generally put a board over the opening as well to make it too low for a cow to squeeze under. If you have an old shed not being used for anything else, it might work well for a creep feeder. We use an old pigsty.

Until you put weaned calves on a finishing ration, grain isn't essential to their diets, but it will help them grow a little faster. Many cattle owners "rough" calves through the winter on hay alone and don't start them on grain until four months or so before slaughter. But feeding weaned calves a pound or two of grain per head per day will quickly teach them to come when called, help them grow, and keep them tame.

FEEDING DAIRY CALVES

Dairy bull calves need a lot of extra care and special feeding for the first few months. While a beef calf gets as much of his mother's milk and affection as he wants for the first six months or more of life, a dairy calf loses his mother and his mother's milk within three days of birth. So be kind to these babies, even though they're often incredibly stubborn. They'll be a little lost and stressed and vulnerable to infection and sickness.

If you buy dairy calves, pick up a sack of milk replacer for each calf from the feed store and a two-quart calf bottle and nipple (per calf) from the farm supply store. Follow the directions on the sack for how often to feed and at what temperature. To mix the milk replacer, fill the bottle half full of warm water, add the right amount of milk replacer, shake well, and add the rest of the water.

A girl bottle feeds one of the farm's calves. Bottle feeding, which is necessary when calves have been taken from their mothers at a very young age, is fun for kids.

At first, it may take a little persuading to get the calf to drink from the bottle. If he won't take the nipple, try backing him into a corner, then straddling him with your legs. Pull his head up and hold it with one hand, stick the nipple in his mouth with the other, and squeeze a little milk onto his tongue. Calves usually catch on pretty quickly. To prevent choking, don't hold the bottle any higher than the height of the calf's shoulder. Calves drink amazingly fast, and the milk will be gone long before their sucking instinct is satisfied.

Calves will try to suck on each other, which isn't a great idea, but they seem to do it anyway. Distract them with calf feed. This is a sweetened grain mix that should be fed free-choice (available at all times) from the time they're a few days old. Take some in your hand and stick it in their mouths after each bottle feeding until they figure out how to eat it themselves. Feed it in a bucket or box attached to the side of their pen, placed high enough that they won't poop in it too often. In case you can't get them outside in a small area with some green grass to nibble once they're a few days old, keep some high-quality hay available for them. They won't do more than play with it for a few days, but they need to get used to it now, while they're still young and open minded about trying new things. It helps their digestive systems mature.

One sack of milk replacer and one to two sacks of calf feed will raise a dairy calf until weaning at eight weeks of age or older. Grain feeding should continue according to the directions on the sack of calf feed, with a gradual transition to

A clean dry place to lie down is the first step toward keeping a calf healthy, whether it's a dairy or a beef calf.

Most beef cattle are finished in large commercial feedlots such as this one, where they are kept in open-air pens and fed a ration precisely formulated for maximum weight gain.

an adult ration as the calf's digestive system matures. Get a calf outside and on pasture as early as possible. You can buy calves in winter, but they're even more susceptible then to pneumonia and scours (diarrhea) so make sure to provide them with a draft-free, deeply bedded pen, and keep it clean. It's also a good idea to give them a good grooming with a cattle brush every day. This mimics the cow's licking and is stimulating and comforting for the calf. A happy calf is more likely to be a healthy calf.

FINISHING RATIONS

There are two approaches to fattening a steer for your freezer. The first is to use time and low-cost inputs, and the second is to speed up the process with a formulated ration fed at a high rate. The first approach usually makes the most sense for small farms because it doesn't require an expensive ration or a separate pen. If you don't have any land for grazing, it's possible to put a weaned calf directly onto a finishing ration, provided it's the right breed and a fast-growing animal. However, the animal will still need plenty of forage-based fiber in its diet. Most calves need some time to mature before they will fatten and are better off on pasture or hay until they're at least a year old.

You also have some choice as to when you send an animal to the processing plant. You can have "baby beef" from a steer as young as a year, although steers are normally kept until they're more mature and have put on some exterior fat. A steer from one of the English breeds can be ready for slaughter as young as sixteen months, while a steer from a continental breed may not finish until it's two years old or more, depending to a large extent on how much grain you feed. In a commercial feedlot, steers are kept in pens and fed as much finishing ration as they will eat. On a small farm, it's usually more practical to keep the steers on pasture and feed them grain once or twice a day. If the pasture is excellent, four or five pounds of corn (usually with a protein supplement) each day will have most steers ready for slaughter in three to four months. Generally, it's a good idea to finish a steer before the age of twenty months to ensure a tender carcass. Please remember, however, that these are just rules of thumb; they are not hard and fast formulas. Finishing cattle is not an exact science.

If the steer is on good pasture, it's helpful if you can time the finishing so that it coincides with the end of the grazing season. Cattle gain weight faster and more cheaply on good pasture than on hay. If the steer is out with a cow herd, he can be trained to come to a separate pen for his ration. To do this, watch where the steer normally is when the herd comes in for the morning drink or grain ration. If he's at the front of the line, close the gate behind him and move him forward into another pen. If he's at the back of the line, close the gate behind the cows and feed the steer with a low bucket in the pasture. If he's in the middle, you'll have to finesse getting the cows ahead and keeping the steer behind for a few days until he figures out to stay behind for his ration.

Unfortunately, most of us can't just walk up to a steer, jam a thumb in his back fat, and know that's he's ready. Two beef-raising friends, Barry and Libby Quinn, told me that you just have to develop an eye for finish. Former extension agent, current friend, and lifelong beef producer Dan Riley told us that a steer is ready for slaughter when you can see the fat around his cod, over his pinbones, and on the rear flank. If a steer has fat around the tail-head, he's close to grading prime; if he has a fat brisket, he's too fat.

Reasons for Grass-Finishing Cattle

Farmers have three reasons for grass-finishing their cattle. *Cost*: grass is decidedly cheaper than grain, which is usually used for finishing. *Health benefits for humans and cattle*: There's a body of research showing that grass-fed beef contains high levels of conjugated linoleic acids (CLAs), which appear to help prevent cancer in humans. In addition, a grass-fed steer's digestive system won't support the killer H151 *E. coli* bacterium; grain-finished cattle lack the CLAs and can harbor *E. coli* due to a lower-acid stomach environment. *Environmental advantages*: Grass-fed beef don't occupy feedlots, which, by nature, have a high concentration of manure that pollutes the air (dust), water (runoff), and soil (too much manure creates nutrient levels that cross the boundary from being fertilizer to being pollution).

Grass-Finishing

Cattle can be finished on grass, but it takes expertise to turn out high-quality beef without grain. If you are interested in grass-finishing, you first will need to buy the right cattle. Short-legged animals from the English breeds are probably your best bet. Second, you will need superb pasture, lush and high enough in protein that it will enable a steer to gain no fewer than 1.7 pounds per day for the last ninety days before slaughter. In most areas, this takes a combination of pastures and planted annual forages plus experienced management. However, any cattle can be raised to maturity on grass and slaughtered for edible beef. The meat may be tough and a little gamy tasting, similar to lean venison, but it will feed you. (See "Reasons for Grass-Finishing Cattle.")

Water

Clean water should be available at all times for your cattle. If the water isn't fresh, they may not drink as much as they should. Tip the water tank a couple of times a season and scrub out the algae. A float-valve, available at farm supply stores, will keep the tank full when you're not around. In below-freezing weather, install a tank heater, also available at farm supply stores, and plan on being there to fill the tank daily because a float-valve freezes up in cold weather.

Unless you run a hose out to the paddocks, your cattle will need to come into the barnyard for a drink. When building your paddocks, create lanes giving your cattle easy access to the barnyard. These can be built with the same portable fencing used for the paddocks and should be about ten feet

The health benefits of grass-finished beef are attracting a lot of consumer interest, but producing good beef on grass alone takes the right breed of cattle and superb management.

A rancher breaks up the ice in the cattle's water trough in the dead of winter. In freezing weather, this may be a regular chore unless you have a water setup with a tank heater.

wide. Gates can be made by tying a gate handle onto the wire and adding a loop at the far gatepost to hook the handle through.

It's not necessary to have water available in the pasture. Cattle will walk a long way to get water, and that's usually good for them and their hooves. Cattle need regular exercise to stay healthy just as much as people do, and like people, they can be a little lazy about it. Even when you're outwintering and there's snow on the ground, cattle will plow through it to get to water and their grain ration. Some producers still rely on snow to water their cattle in the winter, but it's difficult for them to get enough to stay fully hydrated. Eating snow also chills them, and they'll lose weight burning calories to stay warm.

SALT AND MINERALS

Salt is essential to cattle, and the best way to make sure they get enough is to provide free-choice, loose salt in a feeder protected from rain and snow. Buy or build a two-compartment feeder, and put salt in one side and a mineral mix geared for your area in the other side.

Mineral deficiencies were a common cause of disease in cattle in the past. The diseases varied from region to region, depending on what was deficient in the soils. That's why it's important to get a mix formulated for your area of the country. As with salt, it's easier for cattle to get enough minerals when the mix is loose rather than in a block.

A simple tank heater from the farm store keeps water available in winter for our cattle. I run the cord through a piece of PVC pipe in order to prevent the calves from yanking the heater out of the tank.

For steers being finished for slaughter, getting enough minerals and vitamins into their feed is especially important. These can be mixed in the finishing ration according to your feed dealer's directions or fed free-choice in a separate feeder as you would normally do with the cow herd. If fed free-choice, the vitamin and mineral mix should be freshened at least once a week.

CHAPTER FIVE

Handling Beef Cattle

L arge cattle operations have always known that handling cattle effectively on a regular basis requires a properly equipped facility, one that includes pens, a crowding tub, an alley, and a chute with a headgate (as discussed in chapter two). Even a small operation with a handful of cattle can build a scaled-down version of a good facility for a reasonable cost, and the right handling methods can make a bare-bones facility work smoothly.

Over the past several decades, there's been significant rethinking of traditional cattle-handling methods, thanks to the work of such pioneers as University of Colorado professor Temple Grandin and cattle handling and marketing consultant Bud Williams, as well as many others. Their work has focused on understanding the natural instincts and behaviors of cattle and using that knowledge to design systems and handling methods that keep cattle calm and tractable. Temple Grandin's cattle-handling layouts and techniques have beed adopted by many major U.S. processing facilities and innumerable small farms and ranches. Bud Williams's herding methods have set the gold standard for moving and holding cattle in the open, particularly in rangeland grazing.

CATTLE HANDLING FUNDAMENTALS

The most important concepts to understand when handling cattle are flight zone and pressure points. *Flight zone* refers to an imaginary circle around the animal that, when you cross it, causes the animal to move away. By working on the edge of an animal's flight zone, you can gently move it in the direction you want. Because the flight zone varies by the tameness of the herd in general and by each individual animal, and can even be affected by weather conditions, it takes a little finessing to find it. Approach slowly until the cattle begin to move away from you. That's the edge of the flight zone. Go farther, and they'll move faster. Back off, and they'll stop.

Pressure points are specific spots along the edge of the flight zone where, if you stand, you can turn cattle in a different direction. The most important pressure point is at the shoulder. If you're in front of the shoulder on the edge of the flight zone, the cow will turn back. If you're behind it, she'll move ahead. If you're to the side and slightly behind the rear of the animal—a second pressure point—she'll move ahead. If you're directly behind her, where she can't see you, she'll turn around to look at you. If you take a very visible stick along, such as a white step-in fence post or something with a small flag on the end, and you hold it out to the side, you'll look three times as wide to the cattle and they'll turn even more easily.

This is critical information to understand and use when you're trying to move cattle along a fence line and through a gate or to herd a calf back to her mother. After you've spent some time in the pasture playing with these concepts, you can get good enough to sort calves from cows for weaning in a holding pen without having to run them through a chute. You may be able to cut out a pair of cows from the herd when you need to get one in the chute. Never cut out and isolate a single cow. Cattle are herd animals, and they get panicky if made to go off by themselves; they may run over you to get back to their friends. Always work with at least pairs, if not triplets, and you'll have much calmer animals.

When you are vaccinating, ear tagging, sorting, loading, or doing other chores that require moving the animals

Putting gentle pressure on the flight zone, at the correct place, encourages these cattle to move in the right direction.

Accustom your cattle to your presence. Walking through this group of weaned heifers every day at feeding time gentles them.

through the handling facility, there are a few other helpful tips for ensuring things happen smoothly, calmly, and without injuries.

ESTABLISHING A ROUTINE

Cattle love it when the same thing happens at the same time every day. Use this to your advantage by establishing a routine that makes your cattle familiar and comfortable with your handling facility. For example, if you're feeding grain, you could feed it in the holding pen at about seven o'clock every morning. One side of our holding pen is lined with bunker feeders, so our cows like being in the pen. In fact, they're usually waiting there for me at feeding time. When you feed, wear the same cap, carry the same grain bucket, and call with the same call. For the first week or so of establishing a routine, you may need to walk out in the pasture with a grain bucket to where the cattle can see

you and then call to them. With new animals, I've even done a Hansel-and-Gretel routine, leaving a trail of little piles of grain leading back to the pen, where the big feed happens. Cattle catch on to the call and the grain bucket very quickly.

You could put the water tank in the holding pen, preferably on a nice cement pad to eliminate the mud. Leave the gate open so cattle can wander in anytime

Did You Know?

You can avoid getting kicked when working around the back ends of cattle by "tailing" them. Raise the tail straight in the air with one hand, grab it around the base with the other, and keep it pulled toward the head. This keeps the animal's hooves on the ground because kicking with the tail in this position hurts the animal's back. Don't be tentative about tailing; if you don't have a good grip at the base, you can still get kicked.

they're thirsty. Once in a while, after they've had their grain or come up together for a drink, shut the gate behind the herd and make them stand in the pen for an hour or so. The more the cattle are accustomed to being in the pen, the easier it will be to put them there and keep them calmly waiting when you need to work with them. Even if you have only a couple steers for the summer, take the time to train them to be penned. Eventually, you'll have to load them on a truck, and you'll save yourself a lot of aggravation if they're already penned when the truck shows up.

If you're rotationally grazing and switching paddocks regularly, that's another opportunity to teach them to come when they're called. Each time you change paddocks, do it at about the same time of day. Call the cows, then lead them to the next paddock. They'll quickly learn to follow, and it's wonderful to see them romp when they hit the fresh grass.

Of course, the cattle won't know what you're trying to do the first few times you switch paddocks. Wait until they come into the pen for grain or water, then close the gate behind them. While they're getting a drink, go out and close the old paddock gate and open the new one. Then go back, open the pen gate, call the cows, and walk slowly out of the lane to the new paddock, calling as you go. Eventually, they'll follow, and after a few repetitions, they'll learn to follow immediately. They may even try to push past you in the lane. If they do,

spread your arms, and give them a dirty look and a firm "cut it out" so they'll learn you're the leader and they need to mind. If one gets really pushy, rap her on the nose with a stick. This is another important point: remember that cattle have a herd hierarchy, and you should be at the top of it.

If you've got one big pasture, aren't feeding grain, and aren't even around on a daily basis, it's still a good idea to teach the cattle that when you show up and they come to your call, good things will happen. Call them, let them see the bucket, then give them a treat in the holding pen. Use apples, carrots, or dried molasses. At first, just a few cattle will figure it out, but eventually they'll all come. If there's a holdout cow, you can either pen the others and wait till she shows up out of curiosity or pen the others, go out in the pasture, and gently herd her in. She'll usually be quite willing to go where she knows the others went.

After a few months of calling and leading, the herd will usually come when you call and follow when you lead. This minor investment of effort is worth its weight in gold when it comes time to

Did You Know?

Many farmers still use the age-old "come, boss" to call their dairy cows, which linguists believe traces back to the ancient Romans and the Latin *bos* for cow. Think how many thousands of years of routine that represents!

My cattle, recognizing the white bucket I carry as the one I feed them grain in, readily follow me. Train your cattle to recognize that good things such as grain come in certain containers, and they'll follow you almost anywhere.

bring the cattle in for handling. If they aren't accustomed to coming at your call, it means you'll inevitably be out there some nasty cold morning, knowing the veterinarian is due in half an hour and wondering how in heck you'll get them corralled in time. The cattle will quickly sense that you're nervous and react exactly the way you'd expect any prey animal to react to a tensed-up predator type: they'll be at the back end of the pasture before you're through the front gate.

If you want to really get ahead of the game, train your cattle to exit the holding pen through the crowding tub and chute, just as they will on handling days. Close the gate behind the herd when they've entered the pen, and open the chute. I have a board cut to fit over the top of the headgate and hold it open so there's no chance of it slamming shut on a cow. This teaches the cattle that the chute is the way out, and it saves you a lot of trouble getting them into the chute on the day you really need them to go. They'll go by themselves, and they'll think it was their idea!

All in the Timing

Unless you're an early riser, I recommend scheduling paddock switches for the late afternoon. I used to do it first thing in the morning, until Gretel the cow took to standing at the barnyard gate around five o'clock, bellowing for me to hurry up. Since the gate is directly across from our bedroom window and I don't like getting up at the crack of dawn, I soon realized I needed to change the routine. Unfortunately, although Gretel was a good cow and an excellent mother, she never figured out that the routine had changed. She kept bellowing. So I sold her. Sometimes, the best way to deal with a problem cow is to get her off your farm.

GETTING READY FOR HANDLING DAYS

Cattle are pretty specific about what they do and don't like. As mentioned in chapter two, they like firm, nonslippery footing, and they like moving uphill and toward light. They dislike going into dark, unknown spaces; they detest loud noises; and they are afraid of strange objects flapping in the wind or glinting in the sun. They will follow each other around a curve, but they'll balk at going into a chute when they can see the end is blocked. Keep these things in mind when you're getting ready for handling days.

Presumably, your facility is already set up with curves and has good lighting. Now, walk around and make sure there are no soda cans, flapping chains, or glittery, fluttering debris anywhere. Make sure there are no broken boards on the floor of the chute. Oil the gates and levers on the crowding tub, alley, and headgate so they operate silently and smoothly.

Think through what you need to get done and how you're going to do it. If you know ahead of time how you're going to move the cattle through the pens, tub, and chute, it'll be much easier for you to communicate the plan to them. Have a Plan B in case they don't like Plan A. If you need only the calves, for example, or only the open cows, figure out how you can quietly sort off the ones you want and turn out the rest. Decide where you can keep the syringes, needles, and vaccines so they're handy but out of the way of random hooves. Have the ear tags numbered and ready, with extras in case one falls into a fresh cow pie.

Good footing, good lighting, and a few curves in the layout make it easy to move cattle into a crowding tub and alley.

How Not to Handle Cattle

In the movies, cattle are handled with horses, lariats, and a big crew of cowboys. Calves are roped, flipped on their sides, and branded. It's strenuous work, even with little calves, and for some reason filmmakers never do show how the cowboys vaccinate the full-grown cattle—not by flipping them on their sides with a rope, I'm sure. In real life, cattle are handled in all sorts of creative ways. Some owners with small herds will pen cattle against a wall or in a corner with a loose gate; some will crowd them so tightly in a small pen that the animals can't move; and others will somehow get a rope around a cow's neck and tie it up tight to a stout post. What these methods have in common is that they're highly stressful to cattle, fairly unreliable as far as getting every head in a herd treated, and quite dangerous for both cattle and handlers.

HANDLING DAYS

Beef cattle are handled just a few times a year. In most areas, calves are castrated, and often vaccinated and ear tagged, in the spring. If you're using artificial insemination, you'll need to get the cows in the chute whenever they're in heat, and you'll want them to be calm. In the fall, you'll vaccinate the whole herd, and heifers will get their brucellosis vaccinations from the veterinarian. If you haven't ear tagged before, you'll do this then. Two or three weeks later, the calves will be back for a booster vaccination. They're also weaned then or shortly thereafter. In late fall, calves and cull cows are loaded into a truck or a trailer for a ride to the auction barn or a new owner, and fat cattle for your own freezer are loaded and sent to the processor. You may also occasionally have to pen a cow or calf that is sick or injured so you can treat her.

To accomplish these things on a cattle-handling day, you'll need to sort off the cows, calves, or steers you want to work with, then move them into the chute and headgate. You'll catch them one by one in the headgate, where you'll administer vaccines, put on ear tags, or give some other treatment, then release them back to pasture. If you're weaning

Here I'm slowly crowding a pair of yearlings into the alley. Before squeezing your cattle with a gate, make they're facing the right direction.

the calves, you'll need to sort them into separate pens or pastures—one for the cows and the other for their calves—as they come out of the headgate. The veterinarian, or the artificial insemination technician, or the truck for loading cattle may be around on handling day, too, and because cattle are very aware of strangers, this will add to the stress they're already feeling from the change in routine.

Once you have the cattle quietly penned and have given them a half-hour or so to calm down, the quickest way to get the job done is to move slowly. The faster you move, the more agitated the cattle will become, and the tougher it will be to get them into the chute. Don't yell. Research has shown that loud noise, including yelling, is more upsetting to cattle than getting slapped or prodded. Use flight zone and pressure points to move a few cattle at a time from the pen

into the crowding tub. Don't fill the tub more than half full because cattle don't like to be tightly packed. Wait until at least two or three are facing into the chute, then slowly move in the crowding gate. The cattle should start filing into the chute. Sometimes, you'll have to back off on the gate to let a cow turn the right direction. You can wave a hand or herding stick gently in her face or pat a rump to get her to turn, but don't yell and don't hit. Things won't go any faster, and for sure the cattle will be more difficult to work the next time.

Raise the gate at the end of the alley, and let a single animal into the chute. In a perfect world, it will walk up to the headgate with just the right momentum to make the gate close on the animal's neck. In the real world, cattle sometimes come in so gently that the gate doesn't close or so hard you'll worry they've bruised their shoulders. With some of our old cows, I don't bother putting them into the headgate if they're calm enough to stand still while I give an injection.

For the cattle you need in the headgate that aren't cooperating, one helpful trick is to walk quickly from their heads to their rumps; that usually makes them jump ahead. If the animals are reluctant to come down the alley, it may be because they see you standing at the end. Duck out of sight or walk quickly to the rear. Sometimes a pat on the rump or a tap with the stick on their hocks will do the trick. Be patient. Give them a little time. In the end, staying nice and easy will take less time than trying to rush them because too much pressure only makes cattle panicky and balky. A panicky cow may lose her head and try anything to get away—running over a handler, trying to jump a fence that's too high and getting stuck halfway across, or trying to crawl under a gate and mangling the gate.

In the full swing of things, you'll have a cow in the headgate, a couple behind her in the alley, a couple in the crowding tub, and the rest still in the holding pen, waiting their turns. Take care of the first cow, release the headgate, and make sure she gets safely out of your way into another pen or back to the pasture before resetting the headgate for the next patient. Look to see whether the next up is a calf or an adult, and adjust the gate accordingly. With practice, patience, calm, focus, and quiet, you should be able to vaccinate a herd of twenty to thirty animals in well under two hours, without getting any manure splattered on your pants. But wear old pants just in case.

LOADING AND TRANSPORTING CATTLE

Getting cattle on a truck or trailer breaks some of the rules for cattle handling. You're asking them to go into something that's dark and a dead end, and it's an unfamiliar step up to get there. Once again, have some patience. If you are loading into a low trailer, you can load directly from your chute. You can use a pair of loose gates on either side to fill in any gap between the headgate and the trailer gate; if you do this, tie the gates down. If you've got enough cattle to warrant bringing in a semitrailer, you'll need a ramp or some other device to get the cattle up to the level of the truck. Ramps

With his head firmly caught in the headgate, this steer can be safely treated. A simple release mechanism lets him out when you're done.

Advice from the Farm

Handling Cattle

Our experts offer some advice on dealing with cattle in various situations.

Extend Your Reach

"At times, it is helpful to impress the cows with size. I either hold my arms out to my sides, like I learned to do when playing basketball as a kid, or carry my trusty 'cow chasing stick,' which extends my reach to the side. This stick is not used to strike an animal; it just gives the illusion of size. When I get in pretty close proximity to the cows, I start making a kind of *sshshsshsshh* noise. It gets their attention. When I want them to move, I change the noise to more of a *shhooo*, soft and low-pitched. I want them to move quietly, not run."

—JoAnn Pipkorn

Build Strong Gates and Fences

"Have a series of gates and fences of reinforced metal or wood for cattle handling. If they start tearing stuff down, they learn that they can, and you've got trouble. One of the most important things we have now in the industry that we didn't used to have is the self-catching headlock. Have a swing gate on at least one side. It makes you willing to do some of those chores you're supposed to do."

—Rudy Erickson

Have an Exit Strategy

"One thing I always do, and it's something my Dad did, is whenever we have beef cattle out in the field we always have a piece of machinery, like a

wagon, in the field. I tell the boys if something starts coming after you and you can't get under the fence, then get under the machinery or on the other side of it. Basically I keep the boys out of there when we have the bull. And you don't want kids running around in the pasture when the cattle are just getting outside in the spring and they're feeling goofy. You can never trust a bull, and you have to have an exit strategy."

—Randy Janke

Cull Troublemakers

"Don't be afraid to cull cows. If you have a fence jumper, or one that bellows constantly, get rid of it! It's not worth the aggravation. We had a cow that never was quiet that had a nice heifer calf. After that calf was weaned, mama went to the butcher shop. It was much more pleasant around the farm after that!"

—Linda Peterson

are available commercially, or you can build your own. Semitrailers need a lot of room to maneuver, so put the ramp where the driver will be able to turn and back up easily.

If you're moving just a few cattle, get them into the crowding tub well before the truck pulls in. Because the noise will upset them, it's best to have them as far along in the process as possible before they begin to tense up. Once the truck or trailer is in position, begin moving the animals down the chute. The one in front, with the best view of where this is all going, will probably be reluctant. The ones behind may begin pushing the leader along, which is usually OK. Don't rush things. Give the steers time to look at the step.

You will probably need to use a little more persuasion than you would in other handling situations. Have any helpers stand out of sight of the cattle, or try walking quickly from the head of the line to the back. You can slap a rump, but don't yell and don't whip out the electric cattle prods. For one thing, it will cost you money because stressed cattle on their way to auction, the processor, or a buyer have more "shrink," or weight loss, than do unstressed cattle. Michigan veterinarian Ben Bartlett even worked out the math back in 2002. He was selling a truckload of steers, which are sold by the pound, with weight calculated by weighing the truckload. If a steer got upset and pooped before it got on the truck, as happens when one is excited, and the average cow pie weighs seven

These cattle are being calmly loaded for transport.

pounds and is therefore worth about $5, then Dr. Bartlett figured he was losing $300 on a load of sixty steers, just from being in too much of a hurry to get them loaded. Stressed animals won't eat or drink either, so they'll lose more weight in the pens at the auction barn, a second type of shrink. Stressed animals also release a lot of adrenaline into their bloodstream, which darkens and toughens their muscle. These animals are known as dark cutters, and their meat is less valuable.

AFTER HANDLING DAYS

The day after you've worked cattle, go back to the same old routine. You want to erase the bad impressions of the day before with all the good things that will continue to happen if they come when called and follow when led. They won't forget about getting shots or tags, but getting the grain or water and remembering that the quickest exit is through the chute will gradually become uppermost in their minds again. And really, it wasn't that bad. Except for the stop in the headgate and the needle jab, it was all part of the routine.

CHAPTER SIX

Keeping Beef Cattle Healthy

B eef cattle raised outdoors on pasture and hay are naturally hardy animals and tend to have few health problems. When they do become ill or injured, it's important to identify the problem and treat it quickly, before it worsens and causes permanent damage to or kills the animal. As a cattle owner, you have a responsibility to take the necessary steps to prevent problems, to recognize abnormal behavior immediately, and to treat minor health problems or call the veterinarian for conditions beyond your skill. Your cattle health program, then, should have three parts: prevention, diagnosis, and treatment.

PREVENTION

Preventing your cattle from becoming sick or injured is much easier than dealing with a half-ton patient that has no interest in your nursing or a vet's ministrations. Start your "cattle wellness" program with these few simple actions.

BALANCED DIET, EXERCISE, AND SHELTER

Like humans, cattle are healthiest when they get enough exercise and plenty of fresh air, are kept warm and dry in cold weather and cool in hot weather, and eat an adequate diet that includes the necessary vitamins and minerals. As discussed in chapter four, they also need salt and clean water available at all times, although it doesn't hurt them to have to walk a distance in order to get to the water—that can be part of their prescribed exercise program. In addition, maintaining a low-stress environment and a regular routine will contribute a great deal to the overall health of the cattle.

VACCINATIONS

Along with good nutrition, water, and shelter, the best and cheapest insurance against cattle disease is vaccination. Every cattle owner should work with a veterinarian to develop a vaccination program appropriate for the ages and types of his or her cattle and for the region.

Types of Vaccinations

Every program should include calfhood vaccinations and annual booster shots for adult cattle. In the past, combination vaccines were often formulated to address only diseases that were a problem in the region in which the cattle lived. Today, with cattle traveling so often and so far in the United States, it's standard to give all cattle a nine-way vaccine that covers bovine rhinotracheitis, viral diarrhea, parainfluenza, leptospirosis, and several other diseases.

A nine-way does not cover the clostridial diseases, which include tetanus, botulism, anthrax, "wooden tongue" (actinobacillosis), and blackleg. There are vaccines for these, but because they're hard on the animal and those diseases aren't common in cattle in northern Wisconsin, where we live, they aren't usually recommended here. They are recommended, and sometimes required, in many other areas. Check with your veterinarian to determine whether any of the clostridial diseases are common in your area. If so, even if there's no legal requirement, it's worthwhile to vaccinate because these illnesses are difficult to treat and can kill animals in as few as twenty-four hours.

It's also possible to vaccinate against rabies, but again, this usually isn't done if rabies is not a problem in the area.

Brucellosis, also called Bang's disease, causes abortions and fever in cattle

Cattle Diseases and People

A number of cattle diseases are transmissible to people. Tuberculosis can be transmitted through raw milk from infected cows as well as through the air in poorly ventilated barns and sheds. Although tuberculosis is now rare in cattle, it does still occur. Five herds in Minnesota, totaling nearly four thousand head of cattle, were destroyed in 2005 after some of the cattle were found to be infected, and states with recent cases of tuberculosis in cattle or deer are subject to animal transport restrictions.

Brucellosis, or Bang's disease, in cattle is transmitted to people through raw milk or during the delivery of calves from infected cows, in which case it is called undulant fever. Anthrax can be transmitted when people handle infected meat or hides, as can foot-and-mouth disease. Cattle owners can also acquire mange, ringworm, toxoplasmosis, leptospirosis, and tapeworms from their animals, though, thankfully, none of these afflictions is common in people today.

All the same, it's a good idea to keep your cattle vaccinations up to date and your premises clean and well ventilated.

A veterinarian tattoos a young heifer after giving her a brucellosis vaccination. The heifer is held securely in the chute by a headgate and halter.

and undulant fever in people. Undulant fever is all but forgotten now in the United States but is still fairly common in other parts of the world. It's transmitted through contaminated meat and unpasteurized dairy products. Brucellosis vaccination is cheap insurance against abortions in your herd. The shot must be administered by a veterinarian when the heifer is between four and eleven months of age. After giving the vaccination, the veterinarian will attach a metal ear tag to the heifer's ear and officially record the shot.

When to Vaccinate

Calves are often vaccinated for some things soon after birth, but many beef

Calves can be vaccinated shortly after birth, or you can rely on the protection of the mother's colostrum, or first milk. Talk to your vet about what is best for your area.

cattle owners rely on the immunity the babies get from their mothers' colostrum, or first milk, for the first few weeks and wait until it's nearly weaning time—at four to eight months of age— to vaccinate. These older calves should be given booster vaccinations two to four weeks after receiving their first vaccinations. Dairy and orphan calves should be vaccinated by two weeks of age, since they aren't getting ongoing immune protection from their mothers' milk, and again at about six months.

Adult cattle should get an annual booster vaccination. Years ago, cows needed to receive their booster vaccinations before they were bred because some of the vaccines could cause abortions or birth defects. A new generation

of vaccines now allows most cows to be vaccinated after they've been bred, but check with your veterinarian to make sure that these boosters are given at the proper time for your cattle and for your region.

Vaccines need to be kept properly refrigerated and replaced once they pass their expiration dates. They lose potency after those dates. Buy your vaccines from your veterinarian, who will have been careful to keep them cool. Don't mix vaccines unless it says specifically on the label that it's safe to do so.

Injection Methods

Most vaccines are injected into the muscle. This is called an intramuscular (IM) injection, and it's the simplest type to give. Use a 2-inch, 16-gauge needle for adult animals and a 1½-inch, 18-gauge needle for calves. Although IM shots have traditionally been given in the rump, *don't* do this. It tends to make a permanent lesion in some of the most valuable meat. Give injections in the neck. Slap the animal's neck a few times, then pop the needle in quickly and deeply with a firm stroke, and depress the plunger. The animal will probably jump, so don't have your arm between him and any sort of bar, where you could get caught and end up with a broken limb. Change needles often because the tough hides quickly make needles dull, and a dull needle hurts more and is harder to push in than a sharp one.

Another type of injection, the subcutaneous (sub-Q) injection, is given just under the skin and is used for many types of medications. This one takes two hands and a few seconds longer than IM injections do. Making sure your arms aren't in a position in which they might get trapped by a plunging cow, grab a pinch of skin between your thumb and forefinger, use your other hand to quickly push the needle lengthwise into the bottom of the fold, then depress the plunger.

The third type of injection, into a vein (intravenous), is used infrequently and should be done by a veterinarian or someone with experience.

The veterinarian can do all your vaccinating for you, but most cattle owners learn to do it themselves. I use a handy device I found at the farm store that's shaped like a pistol. It contains a big syringe that holds up to ten doses of vaccine. The calibrated trigger delivers exactly the right dose with each injection, and I can vaccinate a whole line of cattle without having to reload the syringe. Change the needle every two or three animals, and always use a new, separate needle to draw vaccine out of the bottle into the syringe so you don't contaminate the vaccine. Syringes can be reused if you clean them carefully with soap and hot water and dry them thoroughly. Needles can be cleaned, sharpened, and reused, but they're cheap and it's generally easier to replace them.

Finally, set up a little table or an apron or some other clean and convenient place with the vaccine bottles, syringes, needles, record book, ear tags, and other equipment you'll need. Having everything handy and organized, but out of the way of the cattle, is easier—and

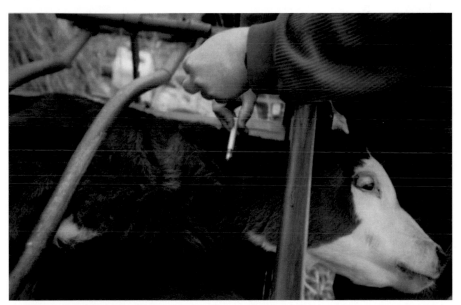

Vaccinating cattle in the neck rather than the rear end, as has been traditional, prevents blemishes in some of the most valuable cuts.

You should have a few basic supplies on hand for treating sick or injured animals. You can add others as the need arises or your veterinarian advises.

The pieces of equipment you definitely want to include in your cattle medical kit:

- **Balling gun**—for getting pills down the throat
- **Rectal thermometer**—when using, put a long string on it; many thermometers have disappeared into cows because of an unexpected muscle contraction

A rope halter often isn't necessary for the basic supply kit, but the rest of the equipment here is: a loaded ear tagger with extra tags and studs, a bottle of 9-way vaccine, syringe, a syringe gun, and a notebook and pencil for recordkeeping.

- **Rope halter and stout rope (couple lengths)**—the halter and rope for holding a head still in the headgate and just the rope for pulling a calf during a tough delivery
- **Stomach tube**—for administering fluids and medications orally and relieving bloat if you're a long way from a veterinarian or are planning on doing a lot of treatment yourself
- **Suturing needles and thread**—for stitching cuts
- **Syringes and needles**—for administering medications

- **Trocar or sharp knife**—for sticking bloated cows

A few medications you'll want to have available that are simple to use and keep well:

- **Baking soda**—for easing stomach upsets in calves
- **Epsom salts**—for digestive upsets and soaking sore or infected feet
- **Iodine**—for treating wounds and for dipping navels on newborn calves
- **Topical antibiotic**—for treating pinkeye and other skin infections

safer—than trying to hold the ear tagger in your teeth and storing the syringe behind your ear.

INTERNAL PARASITES

Many different types of worms like to live inside cattle. The most common parasites are roundworms, lung worms, liver worms, liver flukes, and pinworms. Most cattle owners treat with dewormer medication once or twice a year: in early spring before grazing starts, in late fall after grazing is done for the season, or at both times. For the most effective

Cattle are deloused in the squeeze chute with a pour-on medication. In areas where it's necessary, cattle should be treated for lice and grubs in late fall.

control of worms, discuss with your veterinarian the best time of year to deworm in your area and how to rotate deworming medicines for better results. Worm medicine is widely available at farm supply stores and from your veterinarian. It is either poured along the back or injected.

Concern has risen recently about the increased resistance of internal cattle worms to available medications, and some veterinarians are recommending that cattle not be treated unless worms are really causing a problem. Worm infestation levels are calculated by taking a manure sample and examining it under a microscope for worm eggs, something that can be done only by your veterinarian (although you get to collect the sample). It's easier to first keep a close watch for external signs of a worm problem—weight loss and a dull coat in the summer (lacking the smooth, glossy shine).

Another common internal parasite is cattle grub, the immature stage of heel flies. They live inside cattle until they become adults, then emerge by drilling a hole in the hide of their host. Timing of treatment is crucial to controlling cattle grubs and varies by region. Talk with your veterinarian about your options.

Because internal worms are transmitted by cattle eating grass infected with worm eggs from previous manure deposits, it's possible to reduce worm infestations by managing pasture rotations. Don't return cattle to a paddock until worm eggs deposited in the manure from their last rotation have had time to mature and die. For specific information on the life cycles of different worm species (which vary somewhat by moisture and temperature), consult your veterinarian or university extension service.

Some level of internal parasites is almost inevitable in pastured cattle, but it's usually not a problem unless the infestation is affecting a cow's general health and reducing her resistance to other diseases or slowing weight gain. Gear your treatment to both the nature and the extent of the problem.

EXTERNAL PESTS

Several species of flies delight in tormenting cattle. Cows afflicted with horn flies, face flies, heel flies, or stable flies might run around with their tails in the air trying to get away or quit grazing and bunch together tightly, even on hot days, trying to reduce the amount of hide exposed to bites. When this happens, it's time to do something.

If your cattle are tame and you have time, you can spray them daily with a fly repellent. More economical are medicated ear tags or a repellent-soaked rope or post in the barnyard that the cattle will use for scratching, which you can buy or construct yourself. Inside sheds, you can hang flypaper, long sticky strips that trap flies, or use a light trap or baited trap.

Taking preventive measures to destroy fly-breeding areas will save a lot of money in fly killer. Horn flies lay their eggs in fresh cow pies. If you develop a horn fly problem, try dragging paddocks with a harrow or any sort of homemade drag after grazing to break up the manure pats. Something as simple as an old bedspring behind a car or an ATV (all-terrain vehicle) will work. Stable flies, by contrast, breed best in the mix of old manure and hay that builds up in a ring around round bale feeders. Cleaning up feeding areas in the spring

Cattle spend a lot of energy and grazing time swatting at flies when fly populations are high.

To Their Health

Our experts offer some advice on taking good care of your cattle.

Keep a Close Eye

"How do you spot a sick animal? With my cattle it's really easy—if I take feed out and one is slow coming up, I know something's up. The best way is to check on them every day. I have it a little easier because they're close (to the house). If I walk out to the barn, they're talking to me. I keep a real close eye on them. The meek one, for example, it'll walk up to me and I'll pat it on the forehead and I know it's feeling fine. If it doesn't do that, I know something's up."

—*Randy Janke*

When in Doubt, Call!

"When action is needed, your response may range from catching the animal for closer examination and treatment to calling the vet. When to call the vet is a matter of your comfort level, knowledge, and experience. This is where a knowledgeable and experienced neighbor can be very helpful. Such a neighbor can demonstrate the way to take a temperature or inspect feet or take one look and say, 'Call the vet.' If there is ever a question in your mind about whether to call or not to call a neighbor or vet, call! It's better to look a little overreactive than to lose an animal."

—*Sherri Schulz, DVM*

Set Up a Health Program

"Spotting sick animals is, I think, one of the things that is hard to teach. Check who comes to the feed bunk. If one doesn't come up for feed, it may not be feeling well. If you get a sick one, it'll be standing in the shed or breathing heavily. They recuperate faster if they've had their shots. Another important point is that it's really important to have a vet work with you to lay out a good vaccination program. You don't know all the answers, and it's critical that you get a good product and know how to use it. With some of these new vaccines, if you don't have the herd vaccinated, you shouldn't use it in the calves because it can cause all kinds of side effects."

—*Rudy Erickson*

by removing the detritus will nip a lot of stable fly infestations in the bud. To reduce other flies, in general, keep feeding and watering areas as dry and as free of manure as possible. Some cows seem to have more problems with flies than others (no one knows why). Any cattle that have considerably more fly problems than the rest of your herd should be considered for culling.

Cattle can also suffer from ticks, lice, and mites. Cattle that spend a lot of time scratching themselves on feeders and fence posts probably have lice and should be treated with a spray or powder. Follow the application directions exactly, and don't mix lice treatments with other insecticides. Tick control comes in the form of sprays, dips, and dusts. Scabies mites cause small sores and scabby areas, and mange mites cause thick, wrinkled skin. Roundworm, common in the winter, causes bare patches on the hide. For these problems, talk to your veterinarian about establishing effective control and treatment.

RECORDKEEPING AND ANIMAL IDENTIFICATION

Keeping a notebook to record vaccination, birth, weaning, breeding dates, and other important information for each animal in your herd helps keep your herd's health and breeding programs on time and on track. After a few years of recordkeeping, you'll be able to pick out which animals are healthiest, which breed back the quickest, and which raise the best calves. You'll also be able to cull the cows that are most susceptible to health problems, are poor performers, or have bad dispositions. The result will, I hope, be a herd

Rinderpest

Disease epidemics in cattle—from foot-and-mouth disease to tuberculosis—have a long history of destroying livelihoods and devastating entire countries. The worst of them has probably been rinderpest, a highly contagious viral disease characterized by fever, loss of appetite, bloody diarrhea, and for most of its victims, death. Although it's unknown in the United States, rinderpest is feared throughout Asia and Europe. Rinderpest was brought to Africa in 1889 by cattle imported from India and southern Russia to feed Italian troops. The resulting epidemic destroyed 80 to 90 percent of all cattle in sub-Saharan Africa, as well as killing large numbers of sheep, goats, buffalo, giraffe, eland, antelopes, and wild pigs. The loss of livelihood ruined the social structure of nomadic herding tribes and caused widespread poverty and starvation. The disease returned to Africa in 1982–1984, causing an estimated $500 million in cattle losses. When Dr. Walter Plowright developed a vaccine against rinderpest in 1999, he was awarded the $250,000 World Food Prize. The Food and Agriculture Organization of the United Nations predicts that the vaccine will make it possible to eradicate rinderpest globally by 2010.

Individually numbered ear tags allow you to keep good records of your animals for your breeding program and health concerns.

of uniform, docile, healthy cows that are a pleasure to own and for which you will always have buyers for the calves.

Recordkeeping also plays a crucial role in U.S. animal-health initiatives. Recent outbreaks of foot-and-mouth disease in Britain, of anthrax and tuberculosis in U.S. cattle herds, and of bovine spongiform encephalopathy—mad cow disease—in several countries have fueled a U.S. government initiative to establish a national cattle identification system. The hope is that by identifying each animal individually and tracking the animal's movements, any disease outbreak can be quickly traced to its origins, then suppressed. The United States' scrapie eradication program for sheep already includes an animal identification system, and pig farmers are also moving in that direction. Recently, my state, Wisconsin, established a Premises ID registration as an animal health measure, requiring all livestock owners to register their location and type of livestock owned with the state. Eventually, cattle owners across the country will be required to identify and keep records of each animal in their herds.

For cattle owners who already identify individual animals for their own records and breeding programs, a national animal identification program to control disease outbreaks will require very little extra effort. The standard identification method for cattle is numbered ear tags. The tags and tool for attaching them are available at farm supply stores. Get the large tags; the hairy ears of cattle make it hard to read the little ones. You can buy tags that are already numbered, or you can get a special marker and blank tags and do the numbering yourself. Attach the tag when you've got the animal in the headgate for vaccination, and put it in the center of the ear. Tags too close to the edge have a tendency to rip out. Make sure the tag is on the front side of the ear and the number is facing forward.

Cattle can also be traditionally branded, freeze-branded, or tattooed. These methods require more equipment and are much more painful for the animals. For most small herds, ear tags are more than adequate, with the added advantage of being cheap, quick, and easy to use.

This calf, with its head down and back hunched, looks uncomfortable and may be ill.

HOW TO TELL WHEN AN ANIMAL IS SICK

You can't identify a sick animal until you know how a healthy animal looks and behaves. That's why it's important to spend time on a routine basis just watch-ing your herd. Become familiar with how your cows walk, lie down, get up, chew their cud, lick themselves, stretch, scratch, push each other around, and take care of their calves. This can be a delightful task. When our cows are in the paddock next to the house, there's nothing more pleasant than taking a cup of coffee out on the deck and watching the cows. Summer evenings are especially nice, when it cools off and the calves get frisky. They'll race around the cows, butt heads, and buck. Sometimes the cows come over and watch us.

Once you know what's normal for your cattle, you'll be better able to spot abnormal behavior or appearance. If you get into the habit of regularly taking a good look at each animal, you'll catch problems early. Early diagnosis and treatment is half the battle when it comes to curing a sick or injured animal. A cow that goes off by itself and is not

---

Did You Know?

The fact that cattle appear not to sleep would make it difficult to succeed in the legendary sport of "cow tipping," in which one is supposed to sneak up on a sleeping cow and tip her over. In his 1974 book *Farm Animal Behavior*, Andrew Fraser notes that no concrete evidence exists that adult cattle actually sleep. If they do, it is only for short periods, although they will rest without appearing to sleep for nine to twelve hours a day. Normally, they will rest by standing or lying on their stomachs, although they will lie on their sides for short periods. If you see a cow on its side for longer than an hour, something may be wrong.

Cattle pills are given with a balling gun—a technique you can learn from a vet or an experienced neighbor.

grazing with the others is reason for concern. A cow that is not chewing its cud, spends an abnormal amount of time lying down, has inflamed eyes, shows no interest in its surroundings, or is lame or moves only with effort should be examined more closely. A cow that kicks at its belly, stands still with its back arched, has watery diarrhea, or looks like it's been pumped up with an air compressor needs immediate attention. A calf with watery diarrhea or labored breathing should also be examined immediately.

WHAT TO DO WHEN AN ANIMAL IS SICK

A good veterinarian, a manual of cattle ailments, and a basic medical kit (see "Cattle Medical Kit") are your lines of defense against problems. (Although a list of all the injuries and illnesses that can afflict cattle is beyond the scope of this book, there is a list of the most common ailments in the Appendix.) When you have an animal that is injured or not acting or looking normal and it's not obvious what to do, call your veterinarian and get her advice. You may be able to deal with the problem yourself, or the veterinarian may prefer to make a farm call. If the animal is walking, get him into a clean, bedded pen for observation and treatment or into the chute if that's more appropriate.

As you gain experience with cattle, you'll learn to recognize and to treat minor problems. With good preventive measures and good luck, the only time you'll see your veterinarian is once a year for brucellosis vaccinations and to fine tune your cattle health program.

CHAPTER SEVEN

Breeding Beef Cattle

R aising your own calves is interesting and rewarding. You'll get to pick the parents according to which color, build, temperament, and growth rate you're trying to achieve with your herd. You'll have the pleasure of seeing calves born, watching them grow up, and assessing how well you've done in achieving your goals. Then you get to do it all over again the next year!

CHOOSING COWS AND HEIFERS

A beef cow's primary job is to raise a big healthy calf every year. So the first thing you need to find out about a cow is whether she calves easily and is a good mother.

Easy calving depends as much on the genetics of the bull as it does on those of the cow, so look for this trait when you are evaluating either parent. A heifer's mother should have a record of easy calvings, and reputable bull producers will have records going back several generations that can be correlated with performance records in breed directories. (Ideally, the previous owner will have kept good records.)

If you're looking at a heifer to buy, find out what kind of mother her mother is. A good mother is protective of her calves and diligent about cleaning them up and nursing them quickly after birth. A good mother always knows where her calf is located, and she produces plenty of milk. By weaning time, the calf is chubby, frisky, and a good grazer.

A cow also needs to be easy for you to deal with. A good mother is no good for you if she goes berserk in the chute or is in the habit of jumping fences. Look for cows or heifers that are calm, can be herded quietly, and don't take off for the next county when a stranger enters their pasture.

Get a cow or heifer that can physically handle the job of carrying and birthing calves. She should have wide-set pinbones as well as an udder that is well attached front and back and won't eventually sag so low that the calf will have to kneel to suckle. She should have reached puberty early for her breed, a sign of fertility, and be cycling regularly.

Finally, a cow or heifer should look nice. She should have that beefy, boxcar body, a clean-cut head and neck, a wide muzzle, and calm eyes, and she should not be so fat that she waddles or so thin that her ribs show. Strong, straight legs and tidy hooves are signs she'll hold up for years of grazing and wandering around after her calves.

Of course, the problem with an ideal cow is that she's usually either not for sale or for sale at a price beyond your budget. Do the best you can. Prioritize your wants. My primary concern is to get a heifer that's going to be easy to live with and won't have calving or milk problems. Because I generally work our cattle by myself, I'd rather compromise on body type than on disposition. You'll have to decide for yourself what is important to you.

BREEDING COWS AND HEIFERS

Breed your cows nine to nine and a half months before you want calves. A cow should start cycling, or going into heat, within three weeks to sixty days of having a calf, and she'll cycle about every three weeks after that until she gets bred again. Many cows will breed back (conceive) on their first cycle, while some will take two cycles; those that take three or more or don't get bred at all should be considered for culling.

A heifer is usually bred for the first time the summer after her first birthday, at about fifteen months of age. If a heifer is small for her age, you should

A good-looking group of weaned heifers grazes on pasture.

This "bulling" behavior indicates the heifer is in heat. If the heifer or cow stands still while being mounted, she is in standing heat, ready to be bred.

give her a little longer or cull her. Slow growth is not a desirable trait in beef cattle production.

There are three ways to breed cows and heifers: artificial insemination, a live bull, or a combination of the two.

ARTIFICIAL INSEMINATION

Artificial insemination (AI) allows you to pick from the very best bulls available nationwide, based on detailed statistical information provided by the AI company. In addition to the physical characteristics and growth record of a bull, the company's catalog will list the expected progeny differences (EPD) numbers. This is the statistical chance that a bull's progeny will be above or below the breed average for that characteristic. EPDs are given for such factors as low birth weights (a low birth weight is a strong indicator of easy calving) and weaning weights. Use EPDs to select for characteristics you want to see in your herd, such as better milk production.

Making sense of all the numbers does take some education and practice, so it's a good idea to start by asking advice from your artificial insemination technician. The technician will help you figure out what the numbers mean and which bull is most likely to deliver what you're looking for. If your budget is a little tight, ask about young sires, whose semen is generally cheaper because they aren't yet old enough to have produced enough progeny for a statistically credible EPD record.

The advantages of artificial insemination are that you get a wide choice of excellent bulls and you don't have to worry about keeping a live bull on your farm. If you have just a few cows, it's probably also cheaper than a live bull.

The disadvantage of AI is that you have to spot the cows or heifers when

Heifers should have plenty of good-quality feed their first year, so they will grow well and be ready for breeding in their second summer.

they're in heat, which isn't always easy to do. Heats can be tough to catch on some cows, or you may have an off-farm job and not be around to check cows as often as you need to. To make things easier, you can buy heat-detecting strips. These little bags of dye are glued on a cow's rump. When the cow goes into heat and other cows start trying to mount her, it bursts the bag and mixes the dye with another fluid so that it changes color. This works pretty well, except when you don't get it positioned right or it gets rubbed off, so take care when applying the strips. Attaching the strips to the cow also involves an extra trip through the chute.

A cow cycles every seventeen to twenty-four days, and her heat lasts about thirty-six hours. For best results, she should be bred during the twelve hours she's in a standing heat, which starts about six to twelve hours after the heat cycle begins. A standing heat is so called because during that time a cow or heifer will let another cow mount her. When she's in heat but not in standing heat, the others will smell it and try to mount her, but she won't stand still. But don't worry about the timing of AI breeding too much. As one friend who's bred a lot of cows put it, if you see a cow in heat, breed her. If you're a little late, it's OK.

Did You Know?

Some cattle owners take a class and learn to do artificial insemination (AI) themselves. This allows them to breed a cow in heat without having to wait for an AI technician, which can improve conception rates. Frozen semen is kept on hand in a liquid nitrogen tank. Although AI is not for beginners, you might want to keep it in mind as a possibility for a few years down the road.

To spot cows in heat, you should be out in the pasture at least twice a day. In hot weather, your best chance of catching a heat will be before sunrise, at the coolest time of the day. Once you know a cow is ready to be bred, you have to call your AI technician and then get the cow into a chute. Seventeen days later, start watching the cow carefully to see whether she goes into heat again. If she does, she has to be bred again.

USING A LIVE BULL

The advantage of using a live bull is that the cows and he will take care of the breeding without your needing to get involved. For cattle owners with other jobs or erratic schedules, live breeding saves the headache of scheduling with the AI technician. Cows generally conceive more quickly and at a higher overall rate with a live bull than with AI. If you have enough cows to breed, it will also be cheaper than AI, even if you pay for a high-quality bull.

The disadvantages of using a live bull begin with having to keep a dangerous animal on the farm. On the whole, beef bulls don't have the bad reputation for maiming and killing their owners that dairy bulls do, but you still need to be careful and cautious and to warn visitors.

If you keep the bull for a couple years, you'll need a separate pasture for heifers not old enough to be bred and for cows it's too early in the year to breed. The fences between the pastures must be effective enough to discourage a lovesick bull because bulls can smell a heifer in heat from up to five miles away and will come running if they can.

Another disadvantage of live breeding is that you'll have fewer choices in bull traits than you would with artificial insemination because you'll be limited to whichever bulls are available in your

Keeping a bull separate from the cows and heifers takes a sturdy corral such at this one.

The bull is half the herd, so take the time to look at several before you buy or rent.

area. That said, there are many excellent bull producers, and a bull from your area will be adapted to your climate. Finally, a good bull is expensive and can only be used for two years because after that he would be breeding his daughters.

Choosing a live bull is much the same as choosing cows and heifers, except that your selection has a much greater effect on your herd. A cow has only one calf at a time, but your bull will sire all the calves. The old saying that "the bull is half the herd" is very true. A bad bull can saddle you with poor animals for a long time. Look for a docile animal with a good build. He should be beefy and masculine through the neck and shoulders, with a bit of a hump, and broad and plump from stem to stern, but not fat. A fat bull will get too tired to breed all the cows. A mature bull should be able to breed twenty to twenty-five cows in a season. If you have just a few cows, consider buying a yearling bull, which will be less expensive and, if he's well grown, able to breed up to fifteen cows.

It's also important to use a bull not a lot bigger than your cows. A big bull siring a big calf in a small cow adds up to calving problems.

Instead of buying a bull, you could ask around about renting a bull either before or after the main breeding season, if you're willing to adjust your calving calendar. That way, you'll need to have the bull on your farm for only a couple months.

Another option, if you can find a willing bull producer, is to rent a yearling bull for the summer. Many bull producers like to sell two-year-old bulls, and if you pay them some rent for a yearling for a few months, it cuts their cost of raising a bull to two years. In many areas, you used to be able to get a bull for the summer just for feeding him, but that is happening less and less, and usually not with high-quality animals.

Whatever bull you buy or rent, have him fertility tested. A disturbingly high percentage of beef bulls are infertile, and if you don't catch the problem before you breed, you'll have gone to all that expense and trouble and won't have any calves the next spring—a disaster.

COMBINING AI AND LIVE BULL METHODS

Many serious beef producers use artificial insemination followed by a "cleanup" bull for those cows that don't "settle" with AI. This provides the advantage of getting some top-notch genetics into a herd without having to worry about getting all the cows bred in time to hit your desired calving window. If you've been using AI and having problems getting all the cows to conceive, this might solve the problem.

CARE AND FEEDING OF PREGNANT COWS

A cow that is nurturing a fetus while also nursing a rapidly growing calf works hard and needs a good diet to keep her in condition. A heifer is still growing herself when she is bred for the first time and so needs a good diet, too. A young cow with her first calf and pregnant with the next and that is still growing herself may be working hardest of all.

Fortunately, most beef cows are built to do all these jobs on only good pasture and hay. Even a nursing and pregnant cow shouldn't need any grain, although many producers will supplement their pregnant heifers with a small grain ration through the winter. With some large breeds in certain climates, it may be advisable to supplement grain; for information specific to your cows and region, ask other producers in the area or contact your breed association. Whatever your feeding

These pregnant heifers need adequate feed for continued growth as well as to offset the stress of winter weather. They should be fed as a separate group since their needs for protein are greater than those of mature cows, and it's likely the cows would push the heifers away from the feed.

program, keep an eye on your cows' condition through the winter. If they start getting thin, you'll need to supplement their feed in order to ensure healthy calves the following spring.

CALVING

A clean, green pasture in the spring is a great place for cows to calve. If you're calving in cold or very wet weather, however, you should have a shed or a large pen with a thick layer of clean bedding (straw or old hay) available for the cows. Wherever you calve, plan on checking the cows and heifers several times a day to make sure no one is in trouble. Have a small pen ready for any cow that starts having problems.

As it gets close to calving time, watch the cows' udders closely. Most, but not all, cows will "bag up" before calving. The udder will slowly swell and fill out over a few weeks. Then, a day or so before the calf arrives, the udder will start looking like a balloon that's been blown up to the very edge of popping, and the teats will stick straight out. This is only a general rule, however. Some cows bag up days before the calf arrives, and some don't bag up until the calf is on the ground.

When a cow is ready to calve, bagged up or not, she'll wander off by herself, provided there is sufficient room in the pasture and she can find a quiet corner. She will be restless and will graze just a few bites at a time or quit eating altogether. Some fluid may trickle down her legs, and she may keep her tail cocked. After a while, she may stand still and strain for a minute or so. Sooner or later, she'll lie down and get serious. After some pushing, two little hooves will appear. After some more pushing,

A cow licks her newborn calf. Calves should be born in a clean, dry place; if the cow isn't on grassy pasture, give her lots of clean bedding.

A farmer carefully helps a cow deliver her calf. Calves that are too big or positioned wrong may need your help being born. Gently look and feel until you have an idea what the problem is, then give the appropriate assistance, or call your vet or a neighbor to help.

the nose will peek out. It usually takes a pretty good heave to get the head out; then the rest of the calf will follow in a rush. Immediately after the birth or up to twelve hours later, the placenta is delivered, and the cow usually eats it.

Within minutes of delivery, the cow should be up and licking the calf dry. Within an hour or two, at the most, the calf should be on its feet and looking for the teat. Seeing a calf born, get on its feet, and find milk is an amazing experience. I've seen it quite a few times now, and it's still amazing.

WHEN TO INTERVENE

For the beginning cattle producer, the number one rule is not to intervene during a calving unless it's apparent something is wrong. If a cow's "water" (the fluid-filled membrane surrounding the calf) breaks—you'll see a lot of yellowish fluid—and she hasn't lain down and started hard labor within two to three hours, something is wrong. If a cow has been lying down and straining but fails to deliver a calf within an hour, something is wrong. If you see hooves appear and the entire calf doesn't follow within an hour, then something is wrong.

If, however, the cow is taking longer than expected but is on her feet and puttering around, not in any distress, with no sign of broken water, not straining but taking longer than you expected, chances are that nothing is wrong, and you should just leave her alone. Going out and bothering her will only upset her, slowing down the process even more.

Until you've had some experience with difficult calvings, it's wise to call the veterinarian or an experienced neighbor before attempting to help the cow. Be prepared to give a precise description of the cow's behavior and what you can observe at her back end. If the veterinarian can't come immediately, get some advice about what to do.

If a cow is having a problem but still on her feet, herd her gently and slowly into a pen or chute, where you'll be able to get close enough to give her a hand. If she's lying down, approach slowly, talking softly to her, and try to determine whether the calf is coming out the wrong way or whether the calf is too big for the cow.

A baby that is simply too big for its mama can sometimes be pulled out. Make a loop with a slipknot at the end of two flexible ropes, then slip a loop over each of the calf's feet, just above the first joint of the leg. When the cow begins to strain, start to pull—hard enough to help but not so hard that you risk injuring the calf. When the cow quits straining, quit pulling until she starts again.

If that doesn't bring the calf into the world, get help. If the calf is coming out the wrong way, get help. If the veterinarian can't come, call a neighbor who has some experience. You may end up using a calf puller, or worse, having to kill the calf to save the cow. These are sad situations, and it's why cattle owners pay a lot of attention to getting cows and heifers that calve easily.

If your cows calve out on pasture instead of in a small, closely monitored calving pen, you may not see a cow having her calf, even if you're checking several times a day. As I said earlier, most of them will sneak off to a sheltered corner, often during the night, and in the morning when you check the cattle, you'll find her with a dry, nursing, happy calf. If mother and baby are fine, leave them be and let them rejoin the rest of the herd on their own schedule.

TAKING CARE OF NEWBORN CALVES

There are two schools of thought about what to do with a newborn calf. The first says you should grab that calf as soon after birth as possible, dip its navel in iodine to prevent infection, vaccinate it, ear tag it, and when it's a him, castrate him, then turn him back out with his mother. The second school of thought holds that if a calf is on its feet and nursing and the cow is taking good care of it, leave it alone for a few months. There will be plenty of time to vaccinate, tag, and castrate later. All the immunity it needs is available through the mother's milk.

Each school of thought has its place. If you're calving in freezing cold or wet weather, with plenty of mud around, it's probably a good idea to get right out there and make sure that calf has its navel dipped and gets quickly on its feet and starts nursing. If you've had problems in the past with illness in calves, the extra immune protection from early vaccination may be of real benefit. Putting an ear tag on right away makes it easy to identify cow-calf pairs, especially in a larger herd, and to track which cows are producing the best calves.

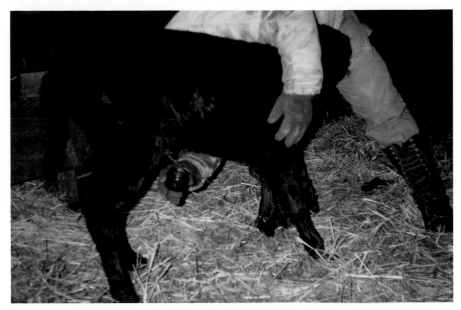

Dipping a newborn calf's navel in tincture of iodine is good preventive medicine.

If you're calving on green grass in warm weather and don't have a history of calf illness, a calf will usually do just fine without the navel dip or the extra vaccination. This saves the calf from the stress of being handled when it's fresh out of the womb, and giving bull calves a few weeks before castration won't hurt them a bit.

CASTRATION

Nobody likes castrating male calves. But the fact of the matter is that many more bull calves are born than will ever be used as breeding animals, and keeping extra bulls around makes no sense if you're interested in calm, easy-to-handle cattle that aren't sexually frustrated and aren't focused on heifers instead of on gaining weight. Get your bull calves castrated before they're four months old.

The younger they are when you do it, the easier it is on them and on you.

There are three ways to castrate bull calves, two of which are not recommended for cattle that will be sold at auction or to a feedlot. The one recommended method is knife castration. It's the bloodiest, quickest, and surest way to get the job done. The best, most effective, and safest way to learn how to perform the procedure is to watch a veterinarian or an experienced owner do it more than once.

Here is description of how the knife castration is performed. Get the calf on his side and tie his legs, and have a helper sit on the calf to hold him down, or put the calf in the headgate and have your helper grab the tail at the base and hold it straight up in the air. (Although when done properly tailing makes it

Castrating can be done by pinning the calf on the ground if you have help—but keep an eye on that cow!

impossible for the animal to kick, I still like to work from the side, just in case.) Grab the scrotal sack in one hand and with a very sharp knife cut off the bottom third. Put the knife aside and encircle the top of the sack with one hand and squeeze until the testicles appear through the opening you've just made. Using your other hand, grab one testicle at a time and pull it away until the connecting tissue and blood vessels break. You can cut these tissues and vessels, but pulling is better because it contracts them and reduces the bleeding. Then apologize to the calf and let him go seek the comfort of his mother.

The other two castration methods are acceptable, although you may lose a little money at the auction barn and there's a greater risk of an incomplete castration. One method involves the use of an elastrator, a plierslike device that puts a rubber elastrator band around the top of the testicles, shutting off the blood supply. The scrotal sac withers and eventually falls off. The other method employs an emasculatome, also similar in appearance to small pliers, which is used to crush the arteries and the spermatic cords that deliver sperm from where it's manufactured in the testicles to where it's stored in the seminal vesicles. Because these methods run a high risk of incomplete castration, a rapidly increasing number of auction barn buyers and feedlots pay a lower price for cattle castrated using either of these methods than they do for the more effective knife castration. A bull calf that still has a working testicle is still half a bull and is called a stag. That's what you get when the rubber band slips or you miss a spermatic cord with the crusher.

DEHORNING

If you're going to dehorn, do it as soon as the horn buds appear. This is best done with an electric dehorner, which is held on the horn bud until it's burned to a crisp. The process is highly painful for the calf, much worse than castration (if the calf's reaction is any indication). I strongly recommend having your veterinarian do this or getting her to teach you how to do this properly because if the growth ring isn't completely destroyed, your calf will grow misshapen horns. If you have polled, or naturally hornless, cattle, you won't have to deal with this chore at all.

WEANING CALVES

Weaning involves separating the cows from the calves so the calves can't nurse anymore and the cows' udders dry up, giving them a rest and a chance to regain a little weight before calving again. It's the most stressful thing that ever happens to a calf, and it's hard on the cows, too. Make it as gentle as

This handy little device is perfect for restraining a calf while dehorning with a battery-powered dehorner.

Advice from the Farm

Birds, Bees, and After
Our experts talk about breeding, calving, and castration.

Bargain Bulls

"I bought a number of bulls from bull test stations (a university or breeder-sponsored program to compare bull weight gains), but I never bought the top bulls. You look at the basic concept behind these test stations: it's a way for purebred breeders to get themselves known. No one is going to consign their fallout bull; they're going to put their best stuff out there. I've bought bargain bulls—paid near the basement price. Only a few of them bring the big money, and most of them sell near the basement price. If the top bull is worth that large amount of money, that bull that's behind him seven or eight places is certainly worth more than you're paying for him. The only fault I've ever heard anyone say about a young bull is he's inexperienced and unproven. I used to use those young bulls for breeding heifers; I'd give them eight or ten heifers to breed. They learned the ropes that way."

—Dave Nesja

Purebred Purchase

"I bought a bull last year for $1,350, a purebred Hereford with papers. I felt fairly confident with him. I might pay a little bit more if I like the temperament than, say, a person who's going to turn him out with a hundred head."

—Randy Janke

Year-Round Bull

"We run the bull year-round with the cows. We start with a new bull and turn it out with the cows in late August or early September. Each year, the calving date is moved up because the bull is with the cows and heifers year-round. Then the fall of the third year, we sell the bull and don't buy a new one till the following fall, and the calving date is back to late spring and early summer, and we start over again moving the calving date back."

—Mike Hanley

Fresh Heifer

"Everyone knows that bulls are dangerous and that a mother cow will protect her calf, but no one told me about heifers in heat. I was in the barnyard one day making a minor fence repair when some of the heifers

wandered over to see what I was doing. This is pretty normal, so I said hello and went on with my work. As I turned away, a big heifer that must have been going into heat tried to mount me. I think they'll mount just about anything when those hormones are flooding. If I had been just a little closer to the fence, I would've been squashed against it and probably not here to tell the story. As it was, I was knocked down but was able to scramble away. So don't turn your back on a bull or a heifer!"

—Ann Hansen

Muddy Babies

"We calve late now, in May. Our farm is muddy, and we calved early for years, and it was an extreme mess. You either have to be ahead of the mud or behind it—if the cows drop the calves in six inches of mud, they're wet and cold."

—Rudy Erickson

Hired Man

"I used to dehorn, do all my own castrating, and it got to the point where it was getting physically very hard on me and it was a lot of stress. I decided I had to reduce the stress. I had the vet out to vaccinate my calves every fall, before weaning, and I had him do the castrating and dehorning at the same time. He was my hired man, and he has insurance and he's a professional. You can't beat that."

—Dave Nesja

Good Accommodations

"The rare times a cow needs help calving, we put the cow in the corral and also put in a round bale and dig out the end of the bale for the calf. The cow and calf have shelter with the bale, plus feed from the bale, easy access to the headgate, and water tank in the corral."

—Mike Hanley

Dry Cows

"Try to put your cows on dry pasture the month you expect them to calve. If you don't, they will pick the muddiest, most manure-filled area possible to calve."

—Linda Peterson

Unpredictable Mother

"A new mother is very unpredictable. She will chase anything, including you and the dog, so never turn your back on her."

—JoAnn Pipkorn

Nutty Cow

"When a cow has her calf, she can be very protective. It's not the time to be sending a ten-year-old kid to check what kind of calf it is. We had a cow that was just nuts when she calved every year, and after three calves we shipped her because we couldn't handle her."

—Rudy Erickson

Kickier Calves

"Castration is easier on them when they're young. We castrate either before they go on pasture in the spring, or when they come in in the fall—they're a little kickier then."

—Rudy Erickson

This group of calves was weaned during winter weather when neighbors' windows were closed. After a few days, they've settled down and adjusted to the change in feed.

possible. The first rule is *not* to wean by loading the calves on a truck and sending them off to the auction barn or feedlot. Although this method is still common, it inevitably results in a lot of sick, and a few dead, calves, and consequently, buyers have learned to pay less for "truck-weaned" calves.

With a small herd on a small farm, it's generally impractical to separate cows and calves by enough distance that they won't be able to hear each other bawling. And they will bawl. It may take as long as three days and nights of steady noise before they either get tired of it or get too hoarse to go on. There's nothing much you can do about it.

The most practical method for the small operator, which also tends to minimize the noise, is called "fence line" weaning. This involves splitting the cows from the calves, then putting the calves back into their familiar pasture and the cows in an adjacent pasture. This at least gives them the comfort of seeing each other, even if the calves can't nurse. If you do this, be aware that a single-wire electric fence will not keep the two groups apart. They need to be on either side of a sturdy, perimeter-type fence, and it doesn't hurt to have an electric wire there, too, as reinforcement. You'll also need to set up a separate water tank and a separate salt and mineral feeder for the calves. There will still be an amazing amount of noise, but the noise doesn't seem to last as long with this method.

Alternatively, you can move the calves into a pen with plenty of hay and water and put the cows back on pasture. The calves may take a little longer to settle down than if they were in their familiar pasture, but they eventually will. If you use this approach, remember that calves should not go directly from pasture to hay. Instead, gradually introduce hay into their diet well before weaning to allow their digestive systems time to adjust. Weaning and a sudden change in diet at the same time are a recipe for illness.

Buyers prefer that calves be weaned for a minimum of three weeks before shipping so plan accordingly. Heifers or steers that you'll be keeping for your own herd need to be kept separate from their mothers for about three months, and heifers should be kept separate from the bull until they're ready to be bred.

REBREEDING COWS

New calves are left with the cows throughout the breeding season. (They don't seem to notice their parents' amorous activities.) Many producers leave the bull with the cows year-round because it results in a much happier bull, and cows generally begin cycling again about the time you'd turn a bull out anyway. Next year's calves may be a little earlier, but you can correct this when you change bulls by not having the new bull delivered until it's a little later in breeding season. If you're using artificial insemi-nation, then the timing of breeding is at your discretion, not a bull's.

As mentioned earlier, a bull that has bred your cows for two years should be sold to prevent him from breeding with his daughters. Some seed stock producers will breed a bull to his daughters, but this is only for the experienced breeder who keeps careful track of his herd's genetic traits. If you really like a bull, you may be able to board and breed the heifers at someone else's farm, perhaps in exchange for doing the same for their heifers, in order to keep him for one more year.

When it's time to sell the bull, you can look for a private buyer, ship him to an auction, or eat him. Bull meat is prized by some, although others think it's a little too flavorful.

Warm Weather Weaning

I try to wean during the first cold snap in fall, when everyone's windows are closed, but it isn't always possible. If you wean in warm weather, when everyone's windows are open, you may not get a lot of sleep that first night, and you risk complaints from nearby neighbors. That's just how it has to be. Fortunately, our neighbors are farmers and retired farmers so the most they ever seem to say is, "I hear you've weaned the calves." At any rate, the noise tapers off after a few days, and the cows and calves settle into the new routine within a couple of weeks.

CHAPTER EIGHT

Marketing and Processing Your Cattle

I t's time to say good-bye. The steers are fat, the calves are weaned, a cow has got-
ten too old to calve, or the bull needs to move on. You have four choices: sell at
auction, sell to a grade and yield processor, have the animal processed locally for
your own freezer or for sale directly from your farm, or sell to a neighbor who wants
stocker calves. With feeder calves, you can also sell directly to a feedlot.

SELLING AT AUCTION

Sending animals to an auction barn for sale is probably the most common way for
small producers to market their beef cattle. An estimated 85 percent of beef cattle
in the United States are sold through a local auction barn at some point in their
lives. These barns hold regular auctions for a variety of different ages and weights
of animals, from special fall feeder calf auctions to weekly auctions of fat cattle and
cull cows.

    To find the auction barns in your region, check your area farm paper, ask the
neighbors, or call your extension agent. If you're fortunate enough to have a
choice of two or three different barns, find out which type of cattle each barn spe-
cializes in, and keep an eye on their sale prices. These prices will be listed, usually
weekly, in farm papers or on the Internet. For example, you don't want to send a
load of beef feeder calves to a barn specializing in dairy cattle because the buyers
obviously won't be too interested in beef calves. If they were, they'd be at the auction barn having the feeder calf sale. Somebody at a dairy barn will probably buy
your feeder calves, but they won't have to bid nearly as high to get them—and
that means you, the owner, lose money.

Once you've decided on an auction barn, call to get more information and instructions for selling your cattle there. Many auction barns have field representatives who will talk to you over the phone or even visit your farm to discuss the best time to sell cattle and how to prepare and transport them to obtain the best price. Ask if you should have the cattle there the day of the auction or the night before. Find out what the availability and charges are for penning, feeding, and watering your cattle before the sale.

Most types of beef cattle don't require any special preparation before an auction, except to make sure they are well watered and fed and are loaded calmly onto the truck. Feeder calves do require some special measures to obtain a top price. Buyers are especially interested in uniform groups of calves, which will grow and finish at the same rate. Usually, buyers will pay a bit more for calves that are already vaccinated and weaned, started on a grain ration (also called bunker broke), and have been knife castrated. Calves that were simply sorted off from their mothers and loaded immediately onto a truck without any of this "preconditioning," as used to be the traditional way of doing it, now commonly receive a lower price than preconditioned calves.

This is a very important point, so I'm going to repeat it: calves that are to be sold as feeders in the fall, whether directly to a feedlot or at auction, should have been weaned a minimum of three weeks, know how to eat grain from a bunker feeder, and have been knife castrated. They also should have

Pens of cattle are auctioned as bidders look on from the catwalks.

Quality Grades for Beef

(From "Quality and Yield Grades for Beef Carcasses," North Central Regional Extension Publication #357, April 1997, by Dennis E. Burson, University of Nebraska Extension Meats Specialist)

Quality grade indicates the expected palatability or eating satisfaction of the meat. USDA beef quality grades are, in descending order, prime, choice, select, standard, commercial,utility, cutter, and canner. Since quality grading is voluntary, not all carcasses are quality graded. Maturity and marbling are the major considerations in beef quality grading.

There are five degrees of maturity: A—nine to thirty months of age; B—thirty to forty-two months (two and a half to three and a half years); C—forty-two to seventy-two months (three and a half to six years); D—seventy-two to ninety-six months (six to eight years); and E—more than ninety-six months. In young beef carcasses, the lean flesh is light cherry red in color and fine in texture. With advancing maturity, the lean becomes progressively darker in color and more coarsely textured. Only carcasses of A and B maturity are eligible to be graded prime, choice, select, or standard. Carcasses of C, D, and E maturity qualify only for commercial and lower grades.

Marbling, the flecks of fat in the lean, is the other major consideration in quality grading. Although it contributes only slightly to meat tenderness, marbling is probably the greatest contributing factor to the palatability traits of juiciness and flavor.

Bullock carcasses are A-maturity carcasses of male animals that exhibit masculine characteristics, such as a noticeable crest over thick shoulders and a prominent "jump muscle" in front of the hip bone. They often display a slightly darker red lean color and a more coarsely textured lean. Bull carcasses (B, C, D, and E maturity) are not quality graded.

Dark cutters are carcasses that produce a lean that is dark red to almost black in color and has a sticky or gummy texture. This condition often results in cattle that have been stressed for a relatively long period of time.

In review, USDA quality grades are a subjective measure of the meat palatability traits of flavor, juiciness, and tenderness. Other factors, such as genetics, processing methods, types of retail cut, and especially cooking methods, also influence meat palatability.

Calves should be ear tagged, vaccinated, and weaned before being sent to auction.

GRADE AND YIELD

Grade and yield buyers are just what the name implies: they pay according to the carcass grade (prime, choice, select, standard, commercial, utility) and the amount of meat it yields. In other words, they pay so much per pound and per grade. This type of operation can be a good place to send cull cows, which often bring a little more sold by grade and yield than they would be if sold at a regular auction barn. Just as with auction barns, asking around is the best way to find grade and yield operators.

GETTING YOUR BEEF PROCESSED

been vaccinated in the neck and booster vaccinated two to four weeks later, again in the neck. Heifers that are to be sold as breeding stock should have their brucellosis vaccination.

Some auction barns now have programs in which the cattle owner or the owner's veterinarian signs a document stating that these procedures have been done. This will be mentioned during the auction and usually affects the sale price. Yet it's not all about money. When you've preconditioned your calves, you know their sale and move to a new home will be much easier on them. Calves that are preconditioned have lower rates of illness and death compared with calves that are not. The former are less stressed and are more likely to settle into their new homes quickly and to gain weight rapidly. Those calves build your reputation among buyers.

Decades ago, every neighborhood where beef cattle were raised had a small-scale processing plant. Today, if you're lucky, there will still be one in your area. If you're not, you may have to drive a ways. These small-scale plants are completely different from the huge meat-packing plants with their assembly-line cattle processing. At a small plant, each animal is handled individually, and the plant owners and workers are often your neighbors.

Find a reputable plant by asking other beef producers their opinions, then pay a visit. Don't go in the morning, when most of the heavy work is being done and the staff may not have time to talk. Instead, head over in midafternoon, when it's less hectic. The plant should look and smell clean. A reputation for making good sausage is a plus. Talk to the manager, who is often also the owner. Most managers are willing to deal with beginners and will take the time to explain the entire process, from loading your steer onto a truck to picking up the meat. Ask how long it is between the time the animal arrives at the plant and when the animal gets slaughtered, and if it's going to be a few hours, whether there will be water available.

Ask about the processing charges. Most plants charge a per-pound processing fee, with extra charges for fancy butchering or sausage-making. If the plant picks up the steer from the farm, then there will also be a trucking charge.

Once you find a processor and make arrangements for your steer, you'll be asked to decide how you want the meat butchered and packaged. Typically, you'll need to tell them how thick to cut steaks, how many to put in a package, how big you want the roasts (three to five pounds is the usual range), and whether you want all the round steaks and ribs or whether you want these ground for hamburger. You can specify that the heart, liver, and tongue be saved or discarded and that the tail be saved, if you're fond of oxtail soup.

There will be a lot of meat. A typical thousand-pound steer with good beef genetics will generally dress out to around 60 percent or better; dairy steers yield somewhat less. Sixty percent means 350-plus pounds of meat, an amount that will cram a large chest freezer right up to the top. If you don't have room in your own freezer, most processing plants will rent you locker storage space for a very reasonable fee.

If there is no processing plant within a reasonable distance from your farm, you could check into on-farm butchering services. These operators

Grading Beef

Beef can be graded by quality or yield, which is confusing. *Quality grading* is what most people think of when they think of grade, and it refers to the eating quality of the meat. *Yield grade* simply refers to the percentage of the total carcass that is usable meat, and it runs from a 1, the highest yield of edible meat, to a 5, the lowest yield. In general, a good-quality beef steer will yield about 40 to 50 percent of its total body weight in usable meat.

Another term you may run across is *carcass weight*, the weight after the animal has been slaughtered and the head, hide, feet, organs, and digestive tract removed. Carcass weight is also called *dressed weight* or *hanging weight*.

Removing the bones and excess fat from the carcass gives you the *cutting yield*, which is expressed as a percentage of the carcass weight, not of the liveweight. For example, let's say a 1,200-pound steer yields a hanging, or carcass, weight of 725 pounds. If the cutting yield is in the 60 to 70 percent range, as it should be for a good beef animal, the amount of meat you'll take home in little white packages will be 440 to 520 pounds. So a 60 percent cutting yield is the same as 40 percent of the liveweight for this animal.

Retail Beef Cuts…From the Meat Case to the Dinner Table

BEEF MADE EASY™

CHUCK ①

Chuck Arm
Pot Roast, ▣
Boneless

Chuck Shoulder
Pot Roast, ▣
Boneless

Chuck Shoulder
Steak, ▣▣
Boneless

Chuck Eye Steak ▣▣

Chuck Top
Blade Steak, ▣▣
Boneless

Chuck Mock
Tender Steak ▣

Chuck Blade
Steak, ▣▣
Boneless

Chuck 7-Bone
Pot Roast ▣

Chuck Short Ribs ▣

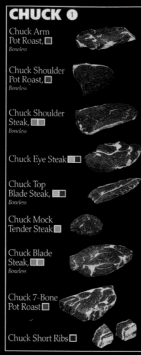

RIB ②

Rib Roast, ▣
Small End, Premium

Rib Steak, ▣
Small End

Ribeye Roast, ▣
Premium

Ribeye Steak ▣▣

Back Ribs ▣

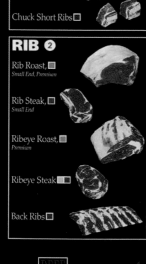

BEEF
IT'S WHAT'S FOR DINNER.
Funded by America's Beef Producers.℠

BEEF

SHORT LOIN ③

Top Loin
(Strip) Steak, ▣▣
Boneless

T-Bone Steak ▣▣

Porterhouse Steak ▣▣

Tenderloin Roast, ▣
Premium

Tenderloin Steaks ▣

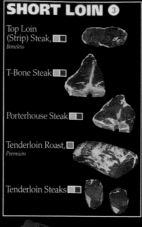

RECOMMENDED COOKING METHOD
▣ SKILLET
▣ GRILL/BROIL
▣ MARINATE & GRILL/MARINATE & BROIL
▣ STIR-FRY
▣ ROAST
▣ STEWING
▣ STEAKS FOR BRAISING
▣ POT ROAST

SHANK ⑥
& BRISKET ⑦

Shank Cross Cut ▣

Brisket, Whole ▣

Brisket, Flat Cut, ▣
Boneless

PLATE ⑧
& FLANK ⑨

Skirt Steak ▣

Flank Steak ▣

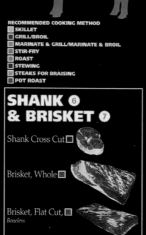

SIRLOIN ④

Top Sirloin Steak ▣▣

Tri-Tip Roast ▣▣

Tri-Tip Steak ▣

ROUND ⑤

Top Round Steak ▣

Round Tip Steak, ▣
Thin Cut

Round Tip Roast ▣

Bottom Round
Roast ▣

Eye Round Roast ▣

Eye Round Steak ▣▣

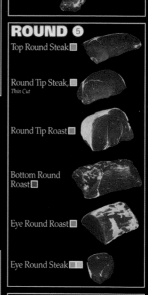

OTHER CUTS

Ground Beef ▣▣▣

Cubed Steak ▣

Beef for Kabobs ▣

Beef for Stew ▣

Beef for Stir-Fry ▣

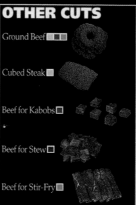

Putting your own beef in the freezer means you'll probably be using a lot of cuts you've never bought before. This chart shows where various cuts come from and how to cook them.

will come to your farm, dispatch your steer, and do the initial skinning and dressing. You may have to do the packaging yourself, or you may be able to pay a little more and have them do it.

SELLING BEEF FROM THE FARM

Selling processed beef directly from your farm is known as direct marketing and used to be how most rural people bought their beef for the year. Traditionally, beef is sold in wholes, halves, and quarters, referring to the proportion of the animal sold to an individual buyer. For example, when a customer buys a half of beef, they get all the meat that came from that half of the steer, from burger to steak. Processors are familiar with this method and will divide up the meat for you if you ask ahead of time. If you're selling quarters, you can sell by front quarter (less valuable), back quarter (more valuable, with lots of steaks), or mixed quarters. With mixed quarters, you take half a carcass and evenly divide the types of meat into two portions. That way, each buyer gets the same amount of steaks, roasts, and burger. There aren't a lot of good steaks on a front quarter.

Some direct marketers sell by the piece, or by marketing each cut individually. This method means you'll probably sell all the steaks and hamburger quickly and have a lot of roasts and round steaks left. It does attract buyers who don't have the freezer space for an entire quarter as well as buyers who would like to try the meat before committing to a larger purchase. If you have a lot of buyers interested in hamburger, you can have cull cows made into burger instead of sending them to a grade and yield operator or to auction. This is terrific burger.

Direct meat sales are regulated by the federal government. It's illegal to sell meat across state lines unless it's been processed in a federally inspected plant. The majority of meat processing plants are state inspected, not federally inspected, so this is a major difficulty for direct marketers living near a state line with a good potential market on the other side. Some have solved this by having their customers drive to the farm to pay for and pick up their meat, while others truck cattle long distances to a federally inspected plant.

Finding buyers for home-grown, grain-fed beef is usually no problem. Relatives, friends, neighbors, and coworkers are all potential customers. But if you sell tough or gamey-tasting beef, you won't get a lot of repeat business. Begin by raising and eating your beef yourself. When you're sure of your beef's quality, then it's time to contact people and ask whether they're interested in buying your beef.

Direct-marketed beef can be priced by liveweight, hanging weight, or dressed weight (see "Grading Beef"). This can be very confusing to customers. The easiest method is to charge by the weight of the meat the customer actually takes home to the freezer. With wholes, halves, and quar-

Beef in a commercial processing plant is inspected and graded by government inspectors.

ters, generally one per-pound price is charged, regardless of the type of meat. This means that on a per-pound basis, the burger is expensive and the steaks cheap; but averaged across the entire purchase, the price is fair. When selling by the piece, the per-pound price varies by whether it's steak, roast, or burger, just as in a grocery store. Many direct marketers add a per-pound processing charge to the price.

Some processors will let your customers pick up their meat directly from the processing plant, while others prefer that you take the meat home; then you can either have your customers pick it up at your place or deliver to them. If you're handling customer deliveries, specify the date and time of pickup or delivery to avoid having big heaps of meat thawing in your garage, waiting to be picked up or delivered to customers. Scheduling this up front enables you to get the meat from the plant to your customers immediately.

TRANSPORT AND SHRINKAGE

Unless you own your own cattle trailer and pick-up truck, you'll need to hire someone to move your cattle from the farm to the auction barn or processing plant. The auction barn or processor will be able to put you in contact with truckers who work in your area, or a neighbor may have a trailer and be willing to do the hauling for you. If you're a member of a beef producers' association and participating in a group sale, trucking arrangements are often included as part of the sale. Some small processors may pick up cattle as part of their services.

Set a date and time for the trucker to show up, and anticipate some slack in the schedule. Truckers often make stops at several farms to fill a trailer, and once in a while somebody has forgotten to pen the cattle or a steer hops a fence, and the timetable gets pushed back an hour or two. Have your animals in the pen, or even in the crowding tub, before the trucker arrives. Most cattle truckers are wizards at backing trailers, but even so, they appreciate it if you make it a straight shot for them to back up to your chute or ramp for loading.

As discussed in chapter five, traveling is stressful for cattle, and they show it by losing weight. Some of the weight is lost in manure or urine, but most is lost by dehydration. This loss is called shrink, and the farther cattle are hauled, the more shrink they will experience. Because most cattle are sold by the pound, shrink reduces their price.

To reduce the amount of shrink, see that your cattle get a good drink before being loaded and that they are loaded calmly. Hiring a calm trucker helps a lot. A trailer with good footing, so they don't slip, also goes a long way toward reducing stress.

Truckers generally charge by miles traveled divided by the number of cattle being hauled. If you're the only customer that day and your cattle fill only half the trailer, your charge per animal will be higher than if you were part of a full load. Trucking charges add up quickly, especially when gas prices are high, so it's a good idea to coordinate cattle shipments with neighbors. Often the trucker will be able to help with this, if you call him far enough in advance. Cattle truckers usually keep a list of people with cattle to ship, and when a trucker's got a full load, he can give you a call and set the date.

Cattle tend to be well-traveled, as they move from their ranch of birth to a backgrounding operation to a feedlot to a processing plant. Cattle trucks like this one are a common sight in many rural areas.

Acknowledgments

I t takes a neighborhood to raise beef cattle. Thanks to everyone who's helped us along the way—the veterinarians, AI technicians, other beef producers, all the guys at Stelter's feed store, our custom hay baler Denny Huse, our trucker Chuck Morning, equipment expert Mike Janke, and the friends and relatives who have encouraged us. I especially want to thank everyone in the Wisconsin Women's Sustainable Farming Network and the producers I've come to know in the Northern Wisconsin Beef Producers Association for their enthusiasm and advice and editor Jim Massey and all the staff at *The Country Today*—my years there as a reporter taught me a lot of what I know about agriculture.

This book wouldn't be what it is without the advice and help I've received from Dan Riley, banker, beef producer, friend, and former extension agent; our neighbor and small-scale beef producer Randy Janke; and all the other producers in the area who've helped out: Prescott Burgh, Rudy Erickson, Donna Foster, the late Mike Hanley, Helen and Bob Kees, Ralph Lentz, Dave Nesja, Linda Peterson, Juliet Tomkins, and many others. Thanks, also, to dairy producer and friend JoAnn Pipkorn and veterinarians Jeff Johnson and Sherri Schulz. And thanks to Des Sikowski-Nelson and her daughter, Val, for spending a cold morning shooting photos at our farm.

Any mistakes, of course, are my own.

Appendix: A Glance at Beef Cattle Afflictions

ACIDOSIS

• Acidosis is a bad stomachache that is caused by a significant and sudden increase in the grain ration for feeder cattle.

• Prevent it by increasing grain rations over a period of a couple of weeks.

• If an animal is kicking at its belly, goes off feed, or shows other signs of distress, call your veterinarian. Untreated acidosis can progress to fever, diarrhea, and laminitis, a dangerous inflammation of the hooves.

BIRTH DEFECTS

• Calves are occasionally born with any number of malformations, from extra toes to crooked legs. Their causes range from genetic defects to the cow's consumption of poisonous plants during her pregnancy.

• Occasionally, a calf will be born with hooves that knuckle under. You can wait for a couple of days to see whether they straighten out. If they don't, the calf may have to be culled. For all other birth defects, consult your veterinarian about the calf's prognosis.

BLOAT

• Bloat is caused when gas builds up in the rumen because a mass of feed is blocking the exit to the esophagus, preventing the animal from belching.

• Bloat usually occurs when cattle are switched too rapidly to a very rich diet, such as to young, lush pasture high in legumes or a lot of grain. Bloat can also occur after a frost or even as the result of heavy dew on rich pasture.

• Bloat occurs rapidly and can kill cattle if the gas is not released. In the early stages, release of the gas may be possible by putting a tube down the animal's throat, but in the late stages it may be necessary to punch a hole in the rumen. If you see one of your herd looking like a swelled-up balloon, call your veterinarian immediately.

COLDS
• Cattle can catch a head cold just as humans do, complete with a cough and runny nose.

• As long as the sick animal is eating well, breathing normally, and not showing any obvious signs of discomfort, the cold can be allowed to run its course. But if symptoms worsen, call your veterinarian. Especially in young calves, a cold can turn into pneumonia.

DIARRHEA (SCOURS)
• Most often seen in young calves, scours is caused either by a bacterial infection or a virus. In older calves or feedlot steers, it's often caused by a coccidiosis (bacterial) infection.

• Knowing the cause is not as important as treating the resulting

dehydration, which is the primary killer of calves with scours. If you see a calf with watery diarrhea, call your veterinarian.

FOOT ROT
• Caused by a bacteria and most common where cattle are kept in muddy or wet areas, foot rot makes cattle lame. The area just above the hoof or between the toes may swell, and often the animal will lose its appetite.

• Untreated foot rot can affect joints and tendons and may cause permanent lameness. Foot rot is treated with antibiotics.

HARDWARE DISEASE
• Cattle often consume some weird stuff along with their hay and pasture, including fencing nails, sticks, bits of wire, and other small metal objects. Usually the junk winds up in the animal's reticulum, the second chamber of the stomach, and causes no harm. Occasionally, a sharp piece of metal will work its way into and even through the wall of the stomach, sometimes as far as the adjacent heart sac and diaphragm. This can cause infection and death.

• Many cattle owners will feed an animal a cow magnet (available at farm supply stores) if they suspect hardware disease, in the hope that any stray metal will stick to the magnet and stay in the stomach. If you notice that an animal has gone off feed and is acting abnormally, hardware disease is a possibility. Call your veterinarian.

LAMINITIS (FOUNDER)

• Laminitis is a severe inflammation of the hoof that can cause the hoof wall to separate from the underlying structure of the hoof and lead to permanent lameness and deformity. Any animal displaying sudden severe lameness should be examined for laminitis.

• The causes of laminitis range from acidosis (see above) to other shocks to the system, such as having to suddenly walk a long way over rough ground or taking too big a drink of cold water.

• Fortunately, laminitis is uncommon in beef cattle, but if a case occurs among your cattle, call your veterinarian.

LUMPY JAW

• The fungus that causes lumpy jaw, actinomycosis, is naturally occurring in the environment and usually does no harm unless it gains entry to an animal's mouth tissues through a cut or sore.

• An animal with lumpy jaw will have a painful swelling on either the upper or lower jaw. As the swelling increases, chewing becomes painful and the animal loses its appetite, which then results in weight loss.

• Treatment of lumpy jaw should be handled by a veterinarian.

MASTITIS

• A bacterial infection of the udder, mastitis causes abnormal-looking milk as well as a painful udder and sometimes fever.

• A nursing cow with an udder that is abnormally swollen on one side or one quarter should be examined for mastitis. Untreated mastitis can turn into a dangerous systemic infection.

• Prevent mastitis by keeping cows on clean bedding or pasture.

NAVEL INFECTION (NAVEL ILL)

• When a newborn calf's navel is contaminated with feces or mud, a bacterial infection can result. Symptoms are swelling around the navel, followed by loss of appetite, diarrhea, and fever over the course of a few days or weeks. The infection may eventually affect the calf's joints or form abscesses in other parts of the body.

• Treatment of navel infection is difficult so prevention is the best policy. See that calves are born on clean bedding or pasture. If calving conditions are less than ideal, navels of newborn calves should be dipped in iodine as soon after birth as possible.

PINKEYE (INFECTIOUS KERATITIS)

• Pinkeye is a contagious and painful eye infection that causes swelling, mattering (discharge), and watering of the eyes. Pinkeye is carried by face flies.

• Most cattle recover from pinkeye without treatment within about three weeks, but some cases can cause permanent eye damage or even blindness. Serious cases can be treated with antibiotics on the advice of your veterinarian.

PNEUMONIA

• Pneumonia is a respiratory disease that often follows a period of high stress or another illness that has lowered an animal's resistance. Symptoms are labored breathing, coughing, fever, and listlessness.

• Pneumonia is a primary killer of young calves, and any calf showing symptoms should be immediately moved to somewhere warm and dry and given plenty of water to drink. Pneumonia affects all ages, and any sign of it should prompt a call to your veterinarian.

POISONING (TOXIC PLANTS)

• A long list of plants found in pastures are poisonous to cattle.

Fortunately, cattle usually won't eat these toxic plants. But if you recently moved your cattle to a new pasture or started feeding them hay from a different source and some of them suddenly take sick, suspect plant poisoning.

• Symptoms of poisoning vary widely, depending on the plant that caused it, and can run the gamut from constipation to diarrhea or from extreme nervousness to total collapse.

• Call your veterinarian if you suspect poisoning.

• Obtain a list of problem plants that are common in your region from your local agricultural extension agent, and make sure you get those plants out of your pastures.

TUBERCULOSIS

• A respiratory disease, tuberculosis is contagious to people, and infected cattle must be destroyed. Once thought to be nearly eradicated in U.S. cattle herds, tuberculosis has recently been found in both Michigan and Minnesota and resulted in the ordered destruction of herds.

• Cattle infected with tuberculosis often show no symptoms at all, but watch for a chronic cough and labored breathing.

- Stay informed of tuberculosis cases in your area through your veterinarian or your extension agent.

WARTS

- Cattle warts are most common on the head, neck, and udder but can appear anywhere on the body.

- Treatment generally isn't necessary or recommended, and most warts will eventually disappear on their own.

Glossary

Afterbirth—the placenta that nourishes the calf during pregnancy; expelled after calving, it is usually eaten by the cow

Artificial insemination (AI)—when a human technician places bull semen in the cow's uterus for impregnation

Auction barn—facility with pens and an auction ring, where cattle and other livestock auctions are held on a routine basis

Aurochs—*Bos primigenius*, the wild ancestor of all cattle, also called wild ox

Background/backgrounder—calves that have been weaned but not yet put "on feed," or a finishing ration

Bag—a cow's udder

Bag up—when the udder swells prior to calving in preparation for milk production

Beef cattle—cattle bred and raised specifically for meat production, as opposed to dairy cattle, which are raised primarily for milk production

Biodiversity—the number of different species in a given habitat, which can include plants, insects, soil organisms, birds, and animals; in a cattle pasture, refers to the number of different plant species present

Bloat—sudden, dangerous swelling of the rumen due to trapped gas (the first chamber of the stomach in cattle)

Body condition score—using a numerical scale to estimate the thinness or fatness of cattle

Booster vaccination—usually, an annual vaccination given to cattle to maintain their immunity against a number of infectious diseases

Bovine spongiform encephalopathy (BSE)/mad cow disease—a slow-moving, always fatal neurological disease of cattle, thought to be transmitted through feed containing parts from contaminated animals

Brand—(noun) a distinctive permanent mark on cattle used to identify them as the property of their owner

Bred—impregnated cow

Breed—(noun) the type of cattle; (verb) to impregnate a cow

Bull—in cattle, an adult uncastrated male

Bulling—when a cow or heifer is in heat and being mounted by other cows or heifers

Calf—cattle less than a year old; male calves are called bull calves, and female calves are called heifer calves

Carcass yield—the percent of the carcass that is edible meat; for example, a 1,000-pound carcass that yields 600 pounds of meat would have a 60 percent carcass yield

Castrate—to remove a male cattle's testicles

Colostrum—the especially rich milk produced by a cow immediately after calving

Condition—the fatness of cattle; a cattle that is losing or gaining weight is said to be losing or gaining condition

Conformation—an animal's shape and build; cattle are said to have good or bad conformation

Commercial herd—producing calves for meat rather than breeding stock; commercial herds commonly contain several breeds and rarely any registered purebred animals

Cow—an adult female bovine that has produced at least one calf

Cow-calf—a type of cattle operation in which a herd of cows is maintained for the purpose of calf production

Creep feed—(verb) to feed calves (usually a grain ration) in an area where cows are kept out by making the entry too small for them

Cud—bolus of food burped up from the rumen to be rechewed in the mouth

Cull—to remove an animal from the herd; in small-scale operations, the most common reasons for culling are poor disposition, age, injury, or illness

Cutting—(verb) to make hay; (noun) a crop of hay; the first cutting is the first crop made in a season; hay is cut from one to ten times a year, depending on the geographic region

Cwt.—abbreviation for hundredweight, or one hundred pounds; cattle prices are often listed by the hundredweight (cwt.)

Dairy cattle—cattle used primarily to produce milk, as opposed to beef cattle, which are raised to produce meat

Dam—the mother of an animal

Dehorn—to remove cattle horns either by cutting them off, burning and killing the horn growth points, or applying a paste to kill the growth points

Dogie—an orphan calf

Dry cow—a cow that is not producing milk; cows are dry from the time their calves are weaned until they produce their next calf

Dual-purpose—cattle used for both milk and meat production; much less common now than in previous centuries

Expected progeny differences (EPD)—a statistical measurement indicating the percentage of a bull's calves likely to be above or below the breed average for that characteristic; EPDs are calculated for birth weight, weaning weight, and several other characteristics

Facilities—the pens, alley, chute, and headgate used for cattle handling; also feeding equipment and barns or sheds used for cattle shelter

Farm—loosely defined as an operation in which crops are raised; in contrast, a ranch raises only livestock and no crops

Feed additives—sweeteners, vitamins, minerals, or medications added to cattle feed to improve palatability and weight gain

Feeder—a bovine on a finishing ration

Feeder calf—a weaned calf either ready to go on feed or, if not yet mature enough, destined to go on hay or pasture until mature enough to go on feed

Feedlot—a cattle operation specializing in finishing cattle for slaughter; feedlot animals are typically kept in pens and fed in bunkers or on cement pads

Finish/finishing—(verb) to fatten cattle for slaughter; (noun) the amount of fat an animal is carrying

Finishing ration—feed used to fatten cattle for slaughter; on a small operation, this is a simple combination of grain and high-quality pasture or hay, while a large commercial feedlot will hire cattle nutritionists to formulate a ration from a wide range of available grains, by-products, and additives

Flight zone—the area around an animal that, if a human enters, will cause the animal to move away

Forage—leaves and stems of plants in the form of pasture or hay

Forbs—nonwoody plants other than grasses or clovers (legumes), such as thistle, goldenrod, and milkweed

Frame—the skeletal size of an animal, usually measured by mature height at the shoulder

Frame size—a numerical scale that defines a mature animal's bone structure and height

Freemartin—a female twin of a male calf; freemartins have a high rate of sterility and should not be bought for breeding

Freeze brand—to brand an animal by using liquid nitrogen

Fresh/to freshen—when a cow gives birth to a calf and begins producing milk

Genetics—animal characteristics determined by genes, such as color and ability to fatten; cattle with desirable characteristics are said to have good genetics

Gestation—the period from the time a cow becomes pregnant to the calf's birth, an average of 285 days in cattle

Grade—a measurement of the tenderness, juiciness, and tastiness of beef; also refers to crossbred (as opposed to purebred) live cattle

Grain—the seeds of plants used as cattle feed, including corn, oats, barley, and rye

Graze—to eat live plants

Hay—cut and dried leaves and stems of grasses and legumes used for animal feed

Heat—the period of a cow's reproductive cycle when she is capable of becoming pregnant

Heifer—a female bovine that has not yet had a calf

Herd—group of cattle

Implant—synthetic hormones placed in the ear of cattle being fattened for slaughter to make them grow faster.

Lactation/lactating—to produce milk

Legume—any of a class of plants that fixes nitrogen in the soil through their roots, including many highly nutritious cattle favorites, such as all clover species and alfalfa

Mad cow disease—*see* bovine spongiform encephalopathy

Management intensive grazing (MIG)—to manage pastures by dividing them into numerous smaller paddocks and rotating cattle through to increase forage production and extend the grazing season; also called rotational grazing

Marbling—the flecks of fat within the lean muscle meat of cattle; a measure of tastiness and juiciness

Open—not pregnant

Organic—food produced in accordance with USDA organic standards prohibiting the use of any synthetic fertilizers, pesticides, and feed additives as well as antibiotics in feed or hormone implants

Ox/oxen—adult, castrated male cattle trained to be harnessed and driven; generations ago, cows were also used for power but were still called cows

Paddock—a fenced-off subdivision of a pasture

Pasture—a fenced-off area used to graze livestock

Placenta—*see* afterbirth

Polled—naturally hornless

Precondition—to wean and vaccinate calves several weeks before sending them to auction

Processor/processing plant—a facility where cattle are slaughtered and butchered

Purebred—an animal whose parents and ancestors are all of the same breed

Ranch—an agricultural operation devoted solely to raising livestock, not crops

Rancher—person who owns or runs (or both) a ranch

Range—unfenced area usable for pasture

Registered—a purebred animal that has been entered in the official registry of a breed organization

Rotational grazing—*see* management intensive grazing

Roughage—the forage component of livestock diet, that is, the hay or pasture

Ruminant—any of a class of animals that have a four-chambered stomach and chew their cud

Ruminate—chewing cud preparatory to digestion

Scur—a malformed horn that grows after an improperly done dehorning

Seed stock—breeding animals

Settle/settled—to become pregnant; to be pregnant

Shrink—weight lost during transport that is due to urination, defecation, or dehydration

Sire—(noun) male parent; (verb) to father a calf

Springer/springing heifer—cow or heifer that is close to calving

Stag—incompletely castrated male

Standing heat—the period during a heat that a cow or heifer will allow a bull to mount and breed her

Steer—castrated male

Stocker—a calf that is being back-grounded, or fed on hay and pasture, until it is mature enough to go on a finishing ration

Stocking rate—the number of animals grazing a pasture, usually calculated in animal units per acre

Straw—the stems of grain plants cut and harvested after they have died at the end of the growing season and the grain has been removed; used for animal bedding and occasionally as animal feed

Veal—calf butchered for meat at a very young age

Wean—to separate a calf from its mother so it can no longer nurse

Weaning weight—a calf's weight at the time of weaning

Yearling—a calf one to two years of age

Resources

It takes a neighborhood to raise beef cattle, plus some good books, magazines, and online resources. For the local information you'll need on seedstock producers, auction barns, cattle haulers, area beef producer organizations, grazing networks, feed stores, custom hay balers, and myriad other necessities, start by chatting with your extension agent. Unless, due to budget cuts, your agent has to cover several counties, he or she is usually headquartered at your county seat. If not, your county offices will be able to tell you where to find the agent who covers your territory. If your agent doesn't know much about beef production or producers' groups in your areas, check at the feed store, with a neighbor, or even with the agricultural reporter for an area newspaper or radio station. Local farm auctions are also good places to meet people and pick up information.

ONLINE RESOURCES

The Internet is a wonderful source of general information on beef cattle and small-scale farming. But use it with caution: much of the information on the Internet is not edited for accuracy and may not apply to your region or farm setup. However, you will find more information on beef cattle faster on the Internet than anywhere else. Some Web sites to get you started on your search for more information are listed here.

BEEF BREEDERS DIRECTORY

Most beef producers don't have Web listings. To find those that do, start with the Web sites listed here. You'll find many more sellers of seedstock or feeder cattle in the want ads in your local paper and in the breeder directories in area agricultural publications.

Breeders World Beef Directory

http://www.breedersworld.com

The beef seedstock producers section is still getting underway in this directory, but there is a good links page to dozens of breed associations, which will direct you to more breeders. There are also listings of sales, events, and fairs.

Bullshop

http://www.bullshop.com

State-by-state listings of bull breeders and sales. Although it's not an extensive listing, it does immediately direct you to your own area.

The Cattle Pages

http://www.cattlepages.com

Easy-to-pick-out listings by location and by breed; the Web site also lists breed associations.

NATIONAL BEEF ORGANIZATIONS

Here are some beef organizations you may want to contact.

Cattlemen's Beef Board

http://www.beefboard.org

Every time you sell a beef animal, one dollar of the price is remitted to the Cattlemen's Beef Board, established by federal order in 1976 to promote beef consumption and the beef industry. The Beef Board uses the money to research and promote consumer awareness of the nutritional benefits of beef and encourage consumption. The Web site contains a variety of information on beef. In addition, you can also call the office in Centennial, Colorado, at 303-220-9890, in order to obtain consumer-oriented information on beef.

National Cattlemen's Beef Association

http://www.beefusa.org

This association of producers, feedlot owners, and meat packers functions primarily as an industry lobbying group, but the Web site has a some useful information on beef statistics, history, and current issues as well as links to affiliated state organizations.

Ranchers-Cattlemen Action Legal Fund United Stockgrowers of America (R-CALF USA)

http://www.r-calfusa.com

A producers-only organization, R-CALF is a comparative newcomer to the national beef cattle scene but has grown rapidly on the strength of its grassroots appeal to small ranchers and its shunning of meat-packer interests. The Web site gives another viewpoint on industry issues, as well as some basic beef information.

RARE BREED CONSERVANCIES

These organizations identify and work to conserve rare breeds of livestock, most of them superbly adapted to very specific regions and environments. If you're interested in helping to preserve cattle breeds that don't fit industry standards but may have much to contribute to diversity and hardiness in cattle, contact these organizations.

American Livestock Breeds Conservancy (ALBC)

http://www.albc-usa.org

The Web site of the ALBC, the largest of the U.S. rare breeds organizations, is very user friendly, and the staff at their headquarters in Pittsboro, North Carolina, is friendly and helpful. Check out their events listings, and read through the list of endangered cattle species.

New England Heritage Breeds Conservancy (NEHBC)

http://www.nehbc.org

This group is committed to helping establish farms focused on developing breeding herds of rare breeds.

PHARMACEUTICALS, EQUIPMENT, AND OTHER BEEF SUPPLIES

If you can't find what you're looking for at your local farm or feed store, check out one of these national suppliers. Regional farm shows are also excellent places to find businesses offering beef equipment and supplies.

American Livestock Supply

http://www.americanlivestock.com
800-356-0700
Vaccines and supplies.

For-Most Livestock Equipment

http://www.for-most.com
800-845-6103
Headgates, chutes, crowding tubs, etc.

Gallagher

http://www.gallagherusa.com
800-531-5908
Electric fencing, watering systems, and scales.

Jeffers Livestock Supply

http://www.jefferslivestock.com/ssc
800-533-3377
Vaccines and supplies.

Kencove Farm Fence Supplies

http://www.kencove.com
800-536-2683
Portable and permanent electric fence equipment and supplies.

Nasco Farm and Ranch Supplies

http://www.eNasco.com/farmandranch
800-558-9595
Everything from ear tags and vaccines to buckets and belt buckles.

Premier 1 Supplies

http://www.premier1supplies.com
800-282-6631
Portable and permanent electric fence equipment and supplies.

Valley Vet Supply

http://www.valleyvet.com
800-419-9524
Vaccines, equipment, and supplies.

W-W Livestock Systems

http://www.wwmanufacturing.com
800-999-1214
Headgates, chutes, scales, etc.

UNIVERSITY RESOURCES

Universities can be excellent places to check for information on beef cattle.

Agricultural Network Information Center

http://www.agnic.org

Fortunately, you don't have to look state-by-state to identify the universities that do the most beef cattle research and extension development. Go directly to this Web site, a collaborative effort between the state extension services and the National Agricultural Library in Beltsville, Maryland. Through this site, you will be able to find university and extension publications on small-scale beef production.

The Cooperative State Research, Education, and Extension Service

http://www.csrees.org

Like the Agricultural Network Information Center, this site identifies which universities do beef cattle research and extension development. In addition, it provides links to local extension offices.

GOVERNMENT RESOURCES

Here are two government Web sites with some great information.

ATTRA (Appropriate Technology Transfer for Rural Areas)

http://www.attra.ncat.org
800-346-9140

ATTRA is a gold mine of free—totally free!—information on whatever topic related to small-scale and sustainable agriculture you can think of. Visit the Web site, and download the publications on small-scale beef production and marketing, or call ATTRA and request a free packet of information tailor-made to answer your questions.

United States National Agricultural Library

http://www.nal.usda.gov

Check out this site on a slow day because you'll need a long time to wade through the wonderfully long list of beef cattle information.

OTHER GOOD SITES

Here are a few other Web sites I recommend visiting for information on beef cattle.

American Veterinary Medical Association (AVMA)

http://www.avma.org

Information on antibiotic use in cattle, mad cow disease, and other health topics.

Cattle Today

http://www.cattletoday.com

Market reports to beef recipes and everything in between.

Cattle Web

http://www.cattleweb.net

Cattle and equipment for sale, market news, and a newsletter; the site is oriented toward larger operations but is still of interest.

Cornell University Poisonous Plants Database

http://www.ansci.cornell.edu/plants/
This extensive listing lets you click on "cattle" to get a list of plants that are of most concern. There are also links to other sites for more regional information.

Grazing Lands Conservation Initiative

http://www.glci.org
A national coalition of organizations and individuals who work to maintain and improve the country's privately owned grazing lands.

Midwest Organic and Sustainable Education Services (MOSES)

http://www.mosesorganic.org
This Wisconsin-based nonprofit organization is a wonderful clearinghouse for information on organic and sustainable agricultural production. MOSES also organizes the largest organic farming conference in the country, the Upper Midwest Organic Farming Conference, every March in La Crosse, Wisconsin. You can also reach them at 715-772-3153.

My Cattle

http://www.mycattle.com
Nice mix of useful information, national and international news, and articles targeted for both large and small producers.

BEEF BOOKS

There are plenty of heavy tomes on beef cattle production aimed at large ranching and feeding operations but not many that deal with the everyday concerns of small-scale beef producers. Below are a few that I have found useful. You may also want to search online used agricultural book sites and used bookstores for out-of-print titles on beef cattle production. Although some of the information in these old books is dated, much of it is more relevant to today's small operator than the college agriculture textbooks now in print.

HANDLING AND FACILITIES

Damerow, Gail. *Fences for Pasture and Garden*. Storey, 1992.
This covers it all, from posts to gates, with plenty of useful information on how to plan and build a variety of fences and gates.

Grandin, Temple. *Beef Cattle Behavior, Handling, and Facilities Design*. Grandin Livestock Handling Systems, 2000.
This book has set the gold standard for laying out livestock facilities and includes an extensive discussion of cattle psychology and behavior and how that knowledge is utilized in designing facilities.

Midwest Plan Service. *Beef Housing and Equipment Handbook*. 4th ed. Iowa State University, 1987.
It's worth the search time if you can find a copy of this out-of-print book. The book is chock-full of plans for barns, sheds, feeding facilities, cattle handling facilities, feed storage bins, waterers,

manure management, fences, and gates. It also adds a lot of useful odds and ends, such as how to put a salt and mineral feeder in a fence line. If this book isn't in your local library or used bookstore, try an Internet search.

RAISING BEEF CATTLE

Haynes, N. Bruce. *Keeping Livestock Healthy: A Veterinary Guide to Horses, Cattle, Pigs, Goats, and Sheep*. 4th ed. Storey, 2001.
An important book if you want to know how to keep your cattle healthy.

Lasater, Laurence M. *The Lasater Philosophy of Cattle Raising*. Santa Cruz Press, 2000.
Tom Lasater established the Beefmaster breed with the goal of combining good genetics, good conservation, and good economics. In this brief, readable book his son, Laurence, describes the Lasater method of cattle production, which offers useful insights for any cattle producer.

Nation, Allan. *Grass Farmers*. Green Park Press, 1993.
Stockman Grass Farmer editor Allan Nation offers a number of case studies of grass-based livestock production in this book, many of them concerning cattle operations. Although the stories are specific to a region and a family, there are plenty of tips here that could be applied anywhere.

Spaulding, C. E., and Jackie Clay. *Veterinary Guide for Animal Owners*. Rodale Books, 2001.
While I haven't read this one, it looks like it has good information on animal care.

Thomas, Heather Smith. *A Guide to Raising Beef Cattle*. Storey, 1998.
This book provides in-depth discussion of all aspects of beef production and is especially helpful with breeding and health questions.

MARKETING BEEF CATTLE

Hamilton, Neil D. *The Legal Guide for Direct Farm Marketing*. Drake University Agricultural Law Center, 1999.
If you're considering selling your beef directly to customers, this book is an essential guide to doing it legally.

Salatin, Joel. *Salad Bar Beef*. Polyface, 1995.
Rotational grazing pioneer Joel Salatin here offers his recipe for successful production and marketing of 100 percent grassfed beef.

Skaggs, Jimmy M. *Prime Cut: Livestock Raising and Meatpacking in the United States, 1607–1983*; and *The Cattle-Trailing Industry: Between Supply and Demand, 1866–1890*. Texas A&M University Press, 1986; University of Oklahoma Press, 1973).
If you're interested learning a little more about the history of beef cattle in the United States, these two histories are enjoyable and very readable.

PERIODICALS

Beef Magazine
http://beef-mag.com
Minneapolis, MN
866-505-7173
This monthly magazine covers the entire beef industry, but the focus is on large producers. Subscriptions are free to operations meeting size guidelines and $35 per year for small producers. *Beef* publishes a special thirteenth issue every February that focuses on cow-calf producers. You can also subscribe to e-mail newsletters, including the *Beef Cow-Calf Weekly*. The related link, http://www.beefcowcalf.com, has links to more than two thousand beef-related fact sheets and Web sites.

Cattle Growers E-Magazine
http://www.cattlegrowers.com
This online magazine offers a plethora of articles and resources on the beef industry.

Hobby Farms
http://www.hobbyfarmsmagazine.com
Lexington, KY 40533
800-627-6157
This bimonthly magazine covers a wide range of informative topics of interest to small-farm owners.

Small Farm Today
http://www.smallfarmtoday.com
3903 W. Ridge Trail Road
Clark, MO 65243
800-633-2535

This monthly magazine established in 1984 is dedicated to promoting small farms and rural lifestyles.

The Stockman Grass Farmer
http://www.stockmangrassfarmer.com
282 Commerce Park Drive
Ridgeland, MS 39157
800-748-9808 or fill in the online form to receive a free sample issue.
If you're interested in grassfed beef or rotational grazing, this is the publication to get. The monthly paper is chock-full of producer profiles, research summaries, and ads. Editor Allan Nation's "Al's Obs" column is a gem.

Index

A

acidosis, 143
additives, feed, 77
AI (artificial insemination), 115–17, 119
alfalfa, 43
alkaloid plants, 47
alleyway, 46
amendments to soil, 71
American breeds, 19
American Livestock Breed Conservancy, 20
anthrax, 100
artificial insemination (AI), 115–17, 119
auction barns, 61, 62, 118, 131–32, 134
aurochs (*Bos primigenius*), 11–12

B

backgrounders, 29, 59
bagged up udder, 120
bales of hay, 44–45, 74, 76
balling gun, 104
barbed wire fences, 33, 34, 35, 36
BCS (body condition scoring), 60
beef, quality grades for, 133, 134, 135
beef cattle (*Bos taurus*)
  biology, 19–22
  breeds, 17–19
  facts about, 39
  history of, 12–16, 17
  reasons for raising, 9, 27, 53
  senses, 23–24
beef cuts chart, 136
Beefmaster breed, 19
beef processing, 134–37
behaviors, 24–26, 28. *See also* personalities
Belted Galloway breed, 17
biology of beef cattle, 19–22, 24

birth defects, 143
Black Angus breed, 18
bloat, 143–44
body condition scoring (BCS), 60
bottle-feeding, 55, 79–80
bovine spongiform encephalopathy, 109
branding, 110
Brangus breed, 19
breeding cattle. *See also* pregnant cows
  age for first time, 114–15
  artificial insemination, 115–17, 119
  booster vaccination warning, 102
  buying cattle for, 54, 59
  calving, 120–22, 127
  calving ease and temperament, 56, 113–14
  cattle production cycle, 28–29
  experts on, 126–27
  gestation, 24
  heat cycle, 24, 114, 115, 116–17
  with live bulls, 117–19
  mastitis, 145
  rebreeding cows, 129
breeds of cattle, 17–19
British breeds, 18–19, 83
brucellosis, 100–101
brucellosis vaccination, 54, 93, 100, 101, 111, 134
Buelingo breed, 17
bull carcasses, 133
bullock carcasses, 133
bulls
  aggressiveness of, 26, 96
  breeding with, 29, 117–19
  buying or renting, 58–59, 118–19, 126
  with cows year-round, 126, 129
  fertility test for, 119
  shape of, 118

bull test stations, 126
butchering services, on-farm, 135, 137
buttercups, 47
buying beef cattle
    bulls, 118
    for calving ease and temperament, 56, 113–14
    deciding what to buy, 53–55, 57–58
    evaluating heifers, 113–14
    experts on, 56
    overview, 53
    pricing and availability, 58–59, 61, 131
    sources for, 61–62

C

calf feed, 80
calves. *See also* weaning
    birth defects, 143
    bottle-feeding, 55, 79–80
    castration, 54, 122, 123–25, 132
    with coccidiosis, 58
    dairy bull calves, 54–55, 62
    dehorning, 125, 127
    feeder calves, 28, 59, 61, 132
    feeding grain to, 78–79, 80–81, 132
    feeding hay to, 80, 129
    and finishing rations, 81
    navel infection, 145–46
    newborn, 122–23
    preconditioned, 132, 134
    scours, 58, 144
    steer calves, 53–54
    truck-weaned, 128, 132
    vaccinations, 101–2, 132, 134
calving. *See* breeding cattle
carcass weight, 135
castration, 54, 122, 123–25, 132
cattle grubs, 105
cattle truckers, 138
Charolais breed, 18, 19
chute with headgate, 46, 48, 91, 94–95
clostridial diseases, 100
coats, 23, 24, 58, 105
coccidiosis, 58
colds, 58, 144
coloring, 17, 18, 24
commercial producers, 61–62
communication, 24
continental breeds, 19
corn, 45–46, 78
corral. *See* holding pen
costs
    of fences, 34
    of handling facility, 49
    of hay and hay feeders, 44, 45
    for purchasing cattle, 59, 61
    up-front nature of, 53
coughing, 58, 144, 146–47
county fairs, 61
cow magnets, 101
creep feeders, 79

crowding pen, 35, 46, 92
culling
    cows that don't breed easily, 28–29, 114
    cows with fly problems, 108
    for small size, 115
    troublemakers, 91, 96
cutting yield, 135

D

dairy-beef cross calves, 55
dairy bull calves, 54, 62
dairy cattle, 17–18, 38, 39, 40
dehorning, 125
Devon breed, 18
deworming, 104–5
Dexter breed, 18
diarrhea (scours), 58, 144
diet changes, 65, 129, 143–44
direct marketing, 137–38
dominance, 25, 26
Dutch Belted breed, 17

E

ear tags, 54, 92, 104, 109–10
elastrator castration, 124
electric fences, 34, 36–37, 62
emasculatome castration, 124
English breeds, 18–19, 83
environmental issues
    benefits of rotational grazing, 70
    creating water access points, 50–51, 69
    erosion from overgrazing, 68–69
    managed versus unmanaged grazing, 66
    pollution from feedlots, 82
    stream banks and erosion, 9, 27
EPD (expected progeny differences), 115
exercise, 99
expected progeny differences (EPD), 115
expenditures. *See* costs
expert advice
    on breeding, 126–27
    on buying beef cattle, 56
    on feed and feeding, 75
    on handling cattle, 96
    on healthy cattle, 107
    on the three Fs, 35
extensive grazing, 67–70
external pests, 106, 108
eyesight, 23

F

facility for handling cattle. *See also* handling cattle; holding pen
    acclimating cattle to, 89, 91
    alleyway, 46
    chute with headgate, 46, 48, 91, 94–95
    crowding pen, 35, 46, 92
    design of, 48–50
    need for, 35
    overview, 46, 48, 87
Farm Animal Behavior (Fraser), 110

Farm Book (Jefferson), 72
fattening cattle
    feeder calves, 28
    in feedlots, 66
    finishing rations, 78, 81–82, 85
    with turnips and carrots, 72
feed and feeding. *See also* pasture
    additives and growth supplements, 77
    bottle-feeding, 55, 79–80
    calves, 78–79, 80–81, 129, 132
    diet changes, 65, 129, 143–44
    experts on, 75
    finishing rations, 78, 81–82, 85
    grain, 38, 45–46, 76, 78–79
    hay, 43–45, 70, 73–76
    legumes, 21–22, 41, 43, 78
    overview, 35, 38, 40
    pregnant cows, 44, 119–20
    salt and minerals, 37, 50, 85
feeder calves, 28, 59, 61, 132
feedlots, 29, 39, 66, 77, 82
fence line weaning, 128
fences
    barbed wire, 33, 34, 35, 36
    building strong, 37, 96, 117
    electric, 34, 36–37, 62
    for fence line weaning, 128
    gates, 38, 85, 96
    overview, 31
    for pastures, 37–38
    portable, 36–37, 70
    removing old fencing, 33–34
    tools for building, 32
fertility test for bulls, 119
finishing rations, 78, 81–82, 85
flies, 63, 74, 106, 108
flight zone, 87–88
fly-breeding areas, 106, 108
foot rot, 144
founder, 145
Fraser, Andrew, 110

G

Galloway breed, 18
gates, 38, 85, 96
Gelbvieh breed, 19
gestation, 24
glycoside plants, 47
Goodnight, Charlie, 14, 16
grain, feeding, 38, 45–46, 76, 89–91
grain, feeding to calves, 78–79, 80–81, 132
Grandin, Temple, 87
grass, ideal care of, 66–67, 71
grass-fed beef, 82, 83
grazing
    keeping records on, 65
    management basics, 70–72
    new cattle, surveillance of, 63
    rotational, 67–70, 90, 105–6
    technique for, 20

as tool for improving land, 9, 27
    unmanaged, 66
growth supplements, 77

H

handling cattle. *See also* facility for
        handling cattle
    experts on, 96
    overview, 87–89
    preparation for, 92, 103–4
    reasons for, 93
    routines, 89–91, 97
    tailing, 89, 123–24
    techniques for, 87–88, 94–95
    for vaccinations, 93, 102–4
hardware disease, 43, 101, 144–45
hauling cattle, 62
hay
    buying and harvesting, 72–73
    feeding, 43–45, 70, 73–76
hay feeders, 45
headgate, 94–95. *See also* chute with headgate
healthy cattle. *See also* vaccinations
    body condition scoring, 60
    experts on, 107
    gait and temperament of, 57
    hooves and hide, 22–23
    preventive maintenance, 99
    signs of, 58
hearing ability, 24
heat cycle, 24, 114, 115, 116–17
heifers, buying or selling, 113–14, 134. *See also*
        breeding cattle
Herd Book of the Shorthorn Breed (Coates), 17
herding instinct, 25, 26, 53, 88
Hereford breed, 18
hides, 23, 24, 58, 105
hierarchy in herd, 25, 26, 90
high-tension fences, 34
holding pen. *See also* facility for handling cattle
    acclimating cattle to, 89–90, 91
    for bulls, 117
    feeding grain and watering in, 89–91
    fences for, 37
hooves, 22–23, 57, 58
hormone implants, 77
human health and grass-fed beef, 82

I

identifying individual animals, 109–10, 122
illness, identifying, 107, 110–11
illnesses and diseases
    anthrax, 100
    birth defects, 143
    bloat, 143–44
    bovine spongiform encephalopathy, 109
    brucellosis, 100–101
    clostridial diseases, 100
    coccidiosis, 58
    colds, 58, 144

coughing, 58, 144, 146–47
foot rot, 144
founder, 145
hardware disease, 43, 101, 144–45
infectious keratitis (pinkeye), 146
internal parasites, 21, 58, 104–6
laminitis (founder), 145
lice and mites, 108
lumpy jaw, 145
mastitis, 145
navel infection, 145–46
pinkeye (infectious keratitis), 146
pneumonia, 58, 146
from poisonous plants, 47, 146
rinderpest, 108
scours, 58, 144
tuberculosis, 146–47
warts, 147
infectious keratitis (pinkeye), 146
information sources, 155–61
intensive grazing, 67–70
internal parasites, 21, 58, 104–6
intramuscular (IM) injections, 102
intravenous injections, 103
iodine, 104

J

Jefferson, Thomas, 72
jimsonweed, 47

K

knife castration, 123–24, 132

L

laminitis (founder), 145
legal requirements for vaccinations, 100
legumes, 21–22, 41, 43, 78
lice treatments, 108
life span, 24
Limousin breed, 19
liveweight, 135, 137
loading cattle, 95, 97. *See also* transporting cattle
Longhorn, 14, 19
lumpy jaw, 145

M

mad cow disease, 109
management intensive grazing, 67–70
mange mites, 108
manure, 21, 58, 144
marbling, 133
marketing cattle, 131–34
mastitis, 145
medical kit, 104
mice and rats, 46
Middle Eastern farmers, 12
mineral mix and salt, 37, 50, 85
mites, 108
moldy hay, 43
monoculture pastures, 72
myths about cattle, 15

N

navel dipping, 122, 123
navel infection, 145–46
nine-way vaccine, 100, 104

O

oats, 45–46
on-farm butchering services, 135, 137
on feed, 28. *See also* feeder calves
outwintering, 76
oxen, 12–13

P

paddocks, 69–70, 83–85, 90, 91
parasites, internal, 21, 58, 104–6
pasture
    caring for, 41, 66–67, 71
    fences for, 37–38
    overview, 40–43, 65
    poisonous plants in, 47
    quality of, 21–22, 63, 66–67
personalities
    behaviors, in general, 24–26, 28
    buying cattle for, 56, 57, 113–14
    herding instinct, 25, 26, 53, 88
    likes and dislikes, 92, 93
    when protecting calves, 126, 127
    when stressed, 97
pinkeye (infectious keratitis), 146
placenta, 121
plants, poisonous, 47, 146
pneumonia, 58, 146
poisonous plants, 47, 146
Pollan, Michael, 77
pollution from feedlots, 82
portable fences, 36–37, 70
post pullers, 33
preconditioned calves, 132, 134
pregnant cows
    care and feeding of, 44, 119–20
    wandering away from herd, 28, 120, 122
    and width of chute, 49
pressure points, 88
pricing processed beef, 137–38
processing plants, 134–35
pulse, 24

Q

quack grass, 42
quality grading, 133, 134, 135

R

rats and mice, 46
recordkeeping, 108–9
rectal thermometer, 104
Red Angus breed, 18
red clover, 41, 43
resources, 155–61
respiration, 24
rinderpest, 108

rodents, 46
rope halters and rope, 104
rotational grazing, 67–70, 90, 105–6
roundworm, 108
routines, 89–91, 97
rub for fly problems, 63, 106
rumen, 20
ruminants, 19–20
running chute, 46, 48

S

Saler breed, 19
salt and mineral feeders, 37, 50, 85
Santa Gertrudis breed, 19
scabies mites, 108
Scottish Highland breed, 18
scours, 58, 144
seed stock producers, 29, 61, 129
selling cattle, 131–34, 135, 137–39
selling processed beef, 137–38
senses, 23–24
sexual maturity, age of, 24
shape of beef cattle, 57–58, 114, 118
shelters, 50, 73, 99, 106, 120
Shorthorn breed, 17, 18
sick cattle, identifying, 107, 110–11. *See also*
    illnesses and diseases
sight, sense of, 23
Simmental breed, 19
sleep habits, 110
smell, sense of, 24, 25
soil testing, 42, 71
squeeze chute, 46, 48, 91, 94–95
stable flies, 74, 106, 107
stag bulls, 125
standing heat, 115, 116
state fairs, 61
steers, 53–54, 82, 83, 85, 135
stocker calves, 28, 59
stomachs, biology of, 20–21
storage bin for grain, 46
subcutaneous (sub-Q) injections, 102–3
suturing needles and thread, 104
syringes and needles, 102–3, 104

T

tailing, avoiding kicks by, 89, 123–24
tank heaters, 83, 84
tattooing, 110
Texas Longhorn, 14, 19
"This Steer's Life" (Pollan), 77
three Fs, experts on, 35. *See also* facility for
    handling cattle; feed and feeding; fences
ticks, 108
tongue, 20
tools for fence building, 32
topical antibiotic, 104
training cattle
    to come when called, 76, 89, 90–91

for electric fence, 37, 62
    to exit the holding pen through the chute, 91
transporting cattle, 62, 95, 97, 138–39
trocar or sharp knife, 104
truck-weaned calves, 128, 132
tuberculosis, 100, 146–47
Types and Market Classes of Live Stock (Vaughn),
    38

U

udders, evaluating, 114, 120
undulant fever, 100–101
U.S. animal-health initiatives, 109
USDA beef quality grades, 133

V

vaccinations
    for brucellosis, 54, 93, 100, 101, 111, 134
    injection methods, 102–4
    for newborns, 122
    overview, 100–101
    schedule for, 93, 101–2
    before selling, 132, 134
Vaughn, Henry W., 38
verbal communication, 24
veterinarians
    calling for advice, 111
    for dehorning, 125, 127
    for vaccinations, 103
    when to call, 107, 121–22

W

warts, 147
water access, 50–51, 69
water consumption, 63
water tank, 50, 83–85, 89–90
weaning
    before buying or selling, 54, 132
    feeding grain, 78–79
    fence line weaning, 128
    overview, 125
    truck-weaned calves, 128, 132
    weather for, 129
weather
    preparation for hot and cold extremes, 50
    shelter from, 50, 73, 99, 106, 120
    for weaning, 129
    winter hay feeding, 73–76
    winter water supply, 83, 84, 85
weed control, 41, 66–67, 71
weight as adults, 24
white snakeroot, 47
Williams, Bud, 87
winter hay feeding, 73–76
winter water supply, 83, 84, 85
wooden fences, 37
wooden posts, removing, 33
worm infestations and medicine, 105

Y

yield grading, 135

ABOUT THE AUTHOR

Ann Larkin Hansen manages the Hansen hobby farm in west-central Wisconsin. Ann has served as a section editor and reporter for the *The Country Today* weekly newspaper and has contributed to publications such as *The Organic Broadcaster* and *Mother Earth News*. She also has authored books in the *The Farm*, the *Farm Animals*, and the *Popular Pet Care* series, published by Abdo and Daughters of Minneapolis.